W9-AVQ-521

Plants
of the
East Bay Parks

Glenn Keator, Ph.D.

Illustrations by
Susan Bazell and
Peg Steunenberg

— *Presented by Mount Diablo Interpretive Association in cooperation with Roberts Rinehart Publishers, Inc.* —

Copyright ©1994 by the Mount Diablo Interpretive Association
Published in the United States of America by Roberts Rinehart Publishers, Inc.
Post Office Box 66, Niwot, Colorado 80544

ISBN 1-879373-42-4
Library of Congress Catalog Card Number 94-66099

Printed in the United States of America

Distributed in the United States and Canada by Publishers Group West

Front cover illustration: Peg Steunenberg
Back cover illustration: Susan Bazell
Book Design: Kowalski Designworks, Inc., Berkeley, California
Designers: Jennifer Crook, Gary Yoshida

In recognition of their generous contribution to this book, I would like to thank the following
photographers who have enriched the book's Photo Gallery. My sincerest thanks go to: Haskel
Bazell, Tim Dallas, Dr. Leon Hunter, and John Karachewski.

Table of Contents

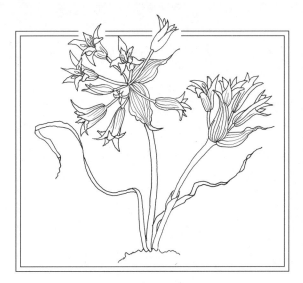

Preface

The original intention to update Mary Bowerman's excellent flora *The Flowering Plants and Ferns of Mount Diablo, California* has expanded to creation of a broader-based guide, aimed at the amateur naturalist, botanist, and biologist. With the expansion in aim comes an expansion in territory. This book covers the two contiguous counties of Alameda and Contra Costa on the eastern side of California's San Francisco Bay. More specifically, the book's aim is to help you identify the plants that are typical of most regional parks in the area. It also is to help you understand why they grow there.

The area covered is bounded by political limits rather than natural ones; this allows a reasonably large number of plants to appear in a field book of manageable size. The species covered are those deemed either most typical or interesting. Though the result is that many species are left out of the plant Encyclopedias, where their traits are detailed, many more species are included in the identification keys and also in the appendix list of plants by location. Other limits are described in the Introduction. Ultimately the species embraced here range through not only the East Bay but adjacent counties around the Bay, and beyond.

We hope that this book finds a useful place in the library of every naturalist, whether visitor or resident. Happy flower hunting!

Glenn Keator

Glenn Keator

Foreword

D r. Keator has organized and condensed an amazing amount of interesting information into a surprisingly small format. In order to make the best use of this valuable information, it is necessary to understand the unusual, but quite logical, organization of the book, which is described in "How to Use This Book."

Dr. Keator's experience giving lectures and short courses to many groups qualifies him to write this book which is readily comprehensible to those without formal botanical training. Although not intended for technical use, professional botanists will most likely discover some new tidbits of information, and enjoy being reminded of forgotten facts. Some statements may stimulate curiosity. For example, is it indeed true that alien plants are not able to grow on serpentine?

One of the challenges confronting the author of a book designed to be useful for people with various backgrounds concerns when to use the scientific or common name of a plant. Dr. Keator has resolved the challenge admirably by providing cross references and, where convenient, utilizing both the scientific and common names.

Another challenge is to make the keys intelligible with limited use of botanical terminology. Dr. Keator has created new artificial keys in which color is used as a primary character. As a result, related species may be separated into different sections, but they are linked to their proper plant family by the key.

A simple description of each family of plants is followed by the names of those members which grow in this geographic area. Subjects discussed may include growth habit and habitat, plant relationships, methods of pollination, historical uses, and origin of the name. Excellent drawings by two gifted illustrators show precise details of the plants. Seventy-five color photos tell the story of the plant communities. The illustrated glossary should be very helpful.

Plants of the East Bay Parks is small enough to travel in your backpack. I am confident that being able to identify plants "on the spot" will enhance the enjoyment of your outings. I hope that Dr. Keator's very readable book will also add to your appreciation of our natural heritage, and encourage you to help preserve representative ecological units of the landscape in our region and beyond.

Mary L. Bowerman

Mary L. Bowerman

Acknowledgments

Plants of the East Bay Parks has been three years in the making. It is available thanks to generous grants from several fine organizations and the efforts of many individuals.

First and foremost, the author gratefully acknowledges the driving force behind the book: the Mount Diablo Interpretive Association (MDIA), a nonprofit volunteer organization affiliated with Mt. Diablo State Park and dedicated to its preservation through education. I particularly acknowledge its executive director, Judy Adler, for the leadership, enthusiasm, and unfailing support that guided the project from conception to completion.

It was a great pleasure to work with two fine artists, Susan Bazell and Peg Steunenberg, who provided detailed, accurate line drawings to accompany the species descriptions. Kevin Hintsa was invaluable in locating specimens for the artists to illustrate.

I am honored that Dr. Mary Bowerman agreed to write the foreword. Her intimate knowledge of Mt. Diablo is an inspiration.

The cooperation of the California Department of Parks and Recreation, especially Superintendents Bob Todd and Larry Ferri, and the East Bay Regional Park District was vital to the project. I would also like to acknowledge the support of the California Native Plant Society and, in particular, Dianne Lake, for reviewing plant names and providing essential plant locations.

I wish to thank those individuals who contributed fine color photographs to our book. To Jennifer Crook and Gary Yoshida of Kowalski Designworks in Berkeley go my thanks for uniting the text and graphic elements into a superlative design.

I am deeply grateful to Wayne Roderick, Toni Fauver, and others who served on the original planning committee for the book; Frank and Edith Valle-Riestra and Celia Kerr for proofreading early drafts of the manuscript; Carey Charlesworth and Patricia Levin for professional copy-editing services; Mary Simpson for fund-raising assistance; Dorothy Lamb, Midge Zischke, Joan Andrews, Myrtle Wolf, and Jake Sigg for recognizing the importance of the book; and Deborah Lee Rose for recommending our publisher.

MDIA would especially like to acknowledge the generosity of the following organizations in providing funding for the book: **The Oakmead Foundation, Orinda Garden Club, and the East Bay Chapter of the California Native Plant Society.**

Introduction

Welcome to the parklands of the East Bay, an area rich in diversity and beauty and full of fascinating trees, shrubs, and wildflowers. Despite the pressures of development by the ever-increasing population, or perhaps because of them, these wild places are treasures to explore, renewing one's spirit of place. This book is dedicated to those places, with the hope of making your experience richer and deeper.

The plants covered in this guidebook are the seed-bearing plants only—conifers and flowering plants—as mentioned, due to space limitations. Ferns, lichens, mosses, liverworts, and horsetails are not included.

The seed plants included here are those that I feel you will be most curious about: the majority of trees, shrubs, vines, and wildflowers you're likely to encounter. The species that have been omitted, including several rare species, those with tiny, inconspicuous flowers such as grasses and sedges, and plants that dwell in marshlands, which at one time encircled the Bay, await another book. Many nonnative, alien species are included to acquaint you with the changes our own civilization has wrought on our wild places. Dandelions, eucalyptus trees, even Monterey pines are out of place, yet seem at home. They have become a part of our story here.

Take this book along whenever you walk the trails of the East Bay Regional Parks or the outstanding centerpiece of our area: Mt. Diablo State Park. This mountain dominates the landscape for scores of miles from whichever direction you approach and beckons you to explore its lofty heights, mysterious canyons, and bountiful woodlands.

How to Use This Book

The book is divided into three main sections and one special middle segment:

I. Plant Communities.

Here learn which plants grow together, why they grow there, and the ways to recognize them. These natural groupings are divided into (1) grasslands (or open places) and their wildflowers, (2) shrublands, also called soft and hard chaparral, and (3) wooded places, which are the forests and woodlands of oaks, other hardwoods, redwoods, and pines.

II. Cross references, Keys and Encyclopedias.

Keys are the only means to accurately identify a new plant you're curious about. Each chapter has its own key, with clear, easy-to-understand language.

For those conversant with common names but not scientific names—or vice versa—each chapter has cross-reference lists with page numbers. In the Encyclopedias, families are arranged alphabetically by common name but genera within each family are in alphabetical order according to scientific name. The cross-reference lists allow you to look up families and genera by the opposite kind of name and locate the entries on them directly, by page number.

After cross-references come the keys. Each of the chapters—on trees, shrubs, vines, and wildflowers—has its own key. Keys are the only means to assure that you're making the correct identification before you turn to the entry describing your plant. Technical terms are kept to a minimum. For added convenience, shrubs and wildflowers are keyed primarily on the basis of their flower shape and color. These groupings allow quick perusal to find the part of the key you need to start with, and minimize the number of steps needed to arrive at the identity of a new flower. Remember that some plants included in the key do not have Encyclopedia entries, simply for lack of space.

In the Encyclopedia, entries have illustrations, brief technical descriptions of identifying traits, and stories about the plants. In cases where a genus is large—for example, the lupines (genus *Lupinus*)—all species are treated under one heading but with brief sketches of common species and their special field characteristics. Over two hundred entries are included.

III. References and Resources.

In Appendix A, organized by park, general locations are given for special plants, with maps. Also featured are the main plant communities that you can find in each park and a discussion of the special status of rare and endangered plants. Appendix B lists the plants of the East Bay, each with some known locations.

Last, turn to the glossary to learn the basics of flower parts, terminology about leaves and their shapes, ways plants are named, and of course various technical words. Most entries are carefully illustrated for easy and clear comprehension. The index guides you by both common and scientific names.

Special Middle Segment.

Photo gallery: plants and plant communities. Interleaved color photos portray selected plants as they look and grow in their communities.

Section I
Plant Communities

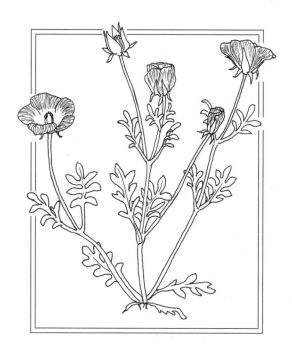

P lant "communities" are useful constructs for describing the wild environments around us; defining a set of community types makes the ever-varying world seem a little more manageable and understandable. Although scientists continue to argue over the vagaries of how communities are defined, we will use simple general categories.

Communities each have personalities of their own determined by their place and time, and the personalities in turn determine what specific plants and animals you find there. The dominant plants (whether they be shrubs, grasses, wildflowers, or trees) give the community its unique character and name. For purposes of practicality, our communities fall into three basic types:

1. Grasslands: open places, without woody plants. Grasslands may represent natural places as well as disturbed, unnatural habitats, such as pastures and grazed areas.

2. Shrublands. Here shrubs occur in dense stands, overshadowing all else. The Spanish term "chaparral"—grove of scrub oaks—has been adopted for many of our shrublands.

3. Wooded places. Trees dominate all such places. There are two general categories: forests (where tree canopies overlap, creating dense shade) and woodlands (where trees are scattered, with openings between).

Chapter 1

Grasslands (Open Places)

Few of California's grasslands and open places have been untouched by humans. Many were, before settlement, quite different communities that were altered deliberately or accidentally, in many cases by removal of shrubs and trees. The most altered open places are where western "civilization" has deliberately created agricultural lands for grazing, production of hay, and nonnative vegetables and fruits, and also for housing tracts. Although few such altered lands are directly considered in this book, many parklands have areas whose past histories reflect such treatments. Many of these are in the process of being reclaimed by native species yet retain an unnatural appearance.

Such open places are what we call "disturbed," meaning there is little natural vegetation. Plowing, tilling, grazing, roadcutting, and burning have so altered the original vegetation that few native plants have survived. Instead these areas are home to plants we call weeds, escapes, and aliens. Weeds are natural here because they're designed to be opportunists, waiting for the chance to move in when land has been cleared. Weeds have evolved special abilities, such as:

- Long-range dispersal of seeds. For example, dandelions float their seeds on hairs, foxtail grasses have sharp barbs that catch in animal fur and human clothing, and bur clover has grappling hooks on its fruits to hitch a ride.
- High viability of seeds.
- Rapid growth in full sun, overtopping and crowding out slower-growing competitors.
- Development of extensive roots that spread rapidly once established, or that delve deeply to gather extra water and nutrients from soils.
- Use of bulblets, taproots, or other underground structures that permit resprouting when tops are cut off.
- Flowering that is prolific, rapid, and without delay under supportive conditions.
- Self-fertility; two individual plants are not required to produce fertile seeds.
- Production of large numbers of seeds.
- Viability of seeds over long periods, so that they germinate only when soils have been tilled, bringing the seeds to the surface where there's light (thus assuring a new crop of weeds every time the soil is turned over).

Open grassland communities exist that are more natural than disturbed. Even before the Europeans arrived, however, grasslands had been altered; Native Americans long practiced burning to keep grasslands open and to exclude woody shrubs and trees. Grasslands were valuable sites of food and basketry plants, and they provided browse for important game animals, such as deer, antelope, and elk.

Our original grasslands were dominated by perennial, native bunchgrasses—clumped grasses that go dormant in summer but do not die. Hundreds of different kinds of annual, perennial, and bulb-bearing wildflowers occur between bunchgrasses, providing a magic carpet of ever-changing color from March through early June. Today, only a few such areas remain to remind us of the original splendor of these grasslands.

The changes that most grasslands have undergone are dramatic. Whether grazed or ungrazed, managed or unmanaged, the majority of grasslands show the effects of the introduction of weedy, nonnative grasses and forbs. Some of these alien grasses and flowers were brought in by design, and others were introduced by accident—often as contaminants of crop seeds (wild oats with cultivated oats, for example), as useful hay crops (alfalfa, sweet clover, red clover), as possible food plants (cardoon, chicory, fennel), or in ballast and bricks.

Native bunchgrasses were generally more palatable than nonnative grasses, so as overgrazing progressed, the demise of these bunchgrasses was inevitable. Bunchgrasses were quickly replaced by annual European and Mid-Eastern grasses, including wild oats (*Avena* spp.), foxtails (*Hordeum* spp.), Italian rye (*Lolium perenne*), bromes (*Bromus* spp.), and fescue (*Festuca* spp.). Meanwhile, flowers with weedy characteristics and long-range dispersal strategies began to fill the spaces between grasses. The greater the grazing pressure, the more the "armed" weeds such as cardoons and thistles took over. Today, much rangeland has been degraded by pernicious, spiny plants like star thistle (*Centaurea solstialis*) and milk thistle (*Silybum marianum*) or by poisonous plants like Klamath weed (*Hypericum perforatum*).

Most grasslands in our parks are at some stage between the extremes of weedlots and natural meadow; many are recovering to bunchgrasses and native wildflowers. But few will ever be completely free of the interlopers. New evidence suggests that light grazing may actually promote better wildflower displays by removing overshadowing grasses at the time of year when wildflowers are actively growing. Much remains to be discovered.

One exception to our grassland story is grasslands on serpentine soils. Serpentine rock—California's slick, soft, shiny bluish-green rock of metamorphic origin—is notorious for its barren, nutrient-poor soils. Serpentine soils are low in essential calcium, high in toxic heavy metals such as molybdenum and nickel, and overly rich in magnesium, a needed nutrient that is nonetheless toxic in large quantities. Consequently, only certain specialized native flowers and grasses, evolving over the eons, have managed to adapt to serpentine soils. Alien weeds and grasses are unable to grow here, so serpentine grasslands give us fine examples of bunchgrasslands in their near-original state.

Meanwhile myriad species of wildflowers, including annuals, summer-dormant perennials, and bulbs, light up our grasslands in spring. Following abundant winter rains and the long, warm days of spring, floral displays explode upon the scene, wherever wildflowers can find a space between grasses. Some years the nonnative grasses get a head start, and wildflowers end up stunted; other years, wildflowers begin growth with or before the grasses and appear in vividly colored masses.

Annuals adapt to California's summer-dry regime by dying when soils dry. Before this, however, they leave behind thousands of summer-dormant seeds. Perennials and bulbs use another ploy: they simply put their extra food and water into safe, underground roots or bulbs until the rains return. Since these subterranean structures are often several inches below the soil surface, they remain cool even during the hottest summers.

Wildflowers belong to numerous families and come in many shapes and sizes, but most are white, blue, purple, or yellow: "bee" colors. (Bee eyes do not perceive orange and red.) Bees are our most abundant, prolific pollinators of open places.

Vernal Pools

Vernal pools have only recently gained the publicity—some would say notoriety—they deserve. Long called hog wallows as well as other degrading names, vernal pools are specialized habitats within our grasslands. Wherever clay soils form small depressions underlain by cementlike hardpans, vernal pools appear. As their name indicates, these are spring pools, filling with water during winter rains and slowly drying as days lengthen and soils warm. Such miniature wetlands are whole ecosystems unto themselves, with special circumstances: seeds, perennial roots, and bulbs must start growth when covered with water, yet they must wait until water levels recede to put on their full growth, blossom, seed, and die. Growing while covered with water is particularly difficult since little oxygen mixes with water, yet plants require oxygen for healthy growth. How vernal pool annuals manage germination under these circumstances is still poorly understood.

The floral displays of vernal pools are impressive indeed. As water evaporates from each level a ring of flowers appears. Most flower species occupy a specific level in their pool, and rings of flowers constantly change. It is not uncommon to see complex swirls and whorls of color like a fine Persian carpet. Whites may blend with yellows, or yellows segue into purples and blues and, lower, mix with more whites.

Vernal pool annuals include downingias, lobelia relatives with perky blue flowers marked and splotched with white, yellow, or dark purple; glue-seed (*Blennosperma nana*), with pale yellow daisy flowers and white pollen; vernal pool mint (*Pogogyne* spp.), with fragrant, mint-scented leaves and bluish two-lipped flowers; popcorn flower (*Plagiobothrys* spp.), with minuscule, white forget-me-not-like flowers; button parsley (*Eryngium* spp.), with spine-edged parsley-scented leaves and tight buttons of spiny-bracted green or bluish flowers; fragrant clover (*Trifolium variegatum*), with small heads of honey-scented white and purple flowers; and annual forms of golden monkeyflower (*Mimulus guttatus*), with perky, golden-yellow two-lipped flowers.

Vernal pools are easily destroyed through habitat degradation, as for example when fields are leveled for agriculture or suburbanization. Since they most often occur in desirable low, rolling foothill country or on valley bottoms, vernal pool habitats are the first to be developed in any given area. Sadly, the East Bay has lost most of its vernal pool habitats. Many were destroyed before the importance of these special wetlands was appreciated.

Chapter 2

Shrublands

Most of our shrublands are called chaparral, but that name actually covers several kinds of shrublands. Generally speaking, chaparral is typical of our hottest, steepest, rockiest slopes—often south facing—where summer sun is strong and winter rains run off (rather than being retained as they are on gentle slopes and flats). We'll detail two broad categories of chaparral: soft and hard.

Soft Chaparral

Also called coastal scrub or coastal sage scrub, soft chaparral is dominated by small shrubs with "soft" leaves (leaves with a pliable, thin texture). Leaves may be heavily scented—smelling of sage, turpentine, or mint—to keep animals from browsing them. These fragrant oils also evaporate on hot days to cool leaves and inhibit the growth of competing plants. All of these ploys prevent shrubs from losing precious leaves, since it costs energy and water to make new ones. Yet in summers with prolonged drought, soft chaparral shrubs may lose most of their leaves as a last-gasp effort to keep from dehydrating faster than roots can replenish water from bone-dry soils. Winter rains bring temporary supplies of water during which leaves are replaced.

Soft chaparral is typical of rocky promontories in the fog belt, but components of this same community appear as temporary replacements for hard chaparral shrubs after brush fires.

Soft chaparral shrubs are varied, with some particularly aggressive pioneer species, such as coyote bush (*Baccharis pilularis*). Others include California sagebrush (*Artemisia californica*), with narrowly divided, sage-scented, gray-green leaves; coffee berry (*Rhamnus californica*), with broad dark green leaves whose edges curl under; sticky monkeyflower (*Diplacus aurantiacus*), with sticky, viscid green lance-shaped leaves, again with curled-under edges; black sage (*Salvia mellifera*), with highly aromatic dark green, narrowly triangular leaves; blue witch (*Solanum umbelliferum*), a green-twigged shrub whose fuzzy, light green leaves are cast away in summer; and poison oak (*Toxicodendron diversilobum*), with shiny, tripartite leaves, which are lost early during severe drought.

Hard Chaparral

Hard chaparral replaces soft chaparral in hotter, drier inland areas, usually on steep, rocky slopes. (Shrubs favor the summer heat of south-facing slopes.) From a distance the dense, tall shrubberies of hard chaparral look like a uniform dark green velvet draped over the mountainsides.

Hard chaparral is so named because its component species have stiff, tough, durable leaves that are seldom shed even at the peak of summer's heat. In fact, the main attribute of such leaves is their long tenancy; shrubs do not have to expend valuable water to create a new set of leaves each year should rains be sparse.

Leaf design varies as much as the several families and genera represented. Manzanitas make stiff ovate leaves that are turned edgewise or vertically to avoid the full brunt of sun—and some kinds, like big-berry manzanita (*Arctostaphylos glauca*), have whitish leaves that reflect away excess light and heat. Chamise (*Adenostoma fasciculatum*) uses narrow, needlelike leaves clustered together to conserve water by minimizing surface area exposed to sun. Wild lilacs (ceanothuses) cover their leaves with a thick, waxy covering that makes them shiny. Bush poppy (*Dendromecon rigida*) has bluish green leaves held obliquely to reflect away heat and minimize the impact of the fierce summer sun.

In addition to their ingeniously designed leaves, chaparral shrubs have deeply probing roots that serve to hold shrubs in place and find sources of deeply hidden water. Roots may also carry on chemical warfare with neighboring shrubs to prevent invasion into their own root zone. Chaparral pea (*Pickeringia montana*) and ceanothuses have tiny knobs on their roots that house nitrogen-fixing bacteria. As a result, such shrubs can move onto nutrient-poor soils; when they die they may pave the way for other shrubs to move in by releasing these nitrogenous compounds into the soil.

Chaparral shrubs grow into nearly impenetrable canopies—from head high to well over ten feet. The best way to pass through is to crawl beneath the branch canopy as small mammals do. Chaparral has been called the elfin forest in allusion to this dense but short forestlike growth pattern.

Chaparral shrubs bloom at varying times on the calendar, although spring favors the most prolific flowering. Water is widely available then, pollinators are most active, and days are warm and long for maximum growth and food production. Manzanitas begin the pageant in late winter and early spring (sometimes as early as the new year); ceanothuses and mountain mahogany begin in early spring; bush poppies, chaparral pea, and coffee berry reveal their blossoms around midspring; and chamise and toyon often herald spring's end in late May to early June. Flower designs, colors, and showiness vary across the spectrum; coffee berry and redberry buckthorn have tiny, yellow-green stars; mountain mahogany lacks petals but has cups of nectar with protruding creamy stamens; chaparral pea has fanciful pea flowers in hot magenta; bush poppy makes large, open yellow saucers filled with numerous pollen-bearing stamens; manzanita hangs racemes of small white or pink urns with minute openings; and ceanothuses mass hundreds of tiny nectar-rich, fragrant white or blue-purple flowers.

Succession in Chaparral

Hard chaparral has long been adapted to the vicissitudes of lightning-caused wild fires. These fires are most likely at the end of the dry summer/fall season. Because fires have long been a frequent ingredient in California's climates, shrubs have adapted by gearing their life cycles accordingly; evidence indicates that before human interference wild fires occurred on the average of every twenty-five to forty years. But because of our fire-suppression policies, we find that old stands of chaparral have become "senescent," literally showing signs of old age through their weakened growth, susceptibility to disease, and dieback of branches. Evidently the leaf duff that accumulates year after year creates natural poisons that weaken shrubs. Without fire, chaparral shrubs eventually die out.

Unfortunately, when fire is forestalled for extended periods fires become extremely hot and destructive because of extra fuel buildup. These fires may be more destructive overall than they are helpful in facilitating regrowth.

With the natural fire cycle, chaparral shrubs renew themselves in two ways: 1) through stump sprouts from long-dormant buds in woody bases, called burls or rootcrowns (if the fire is not so hot that dormant buds are killed); and 2) through seed germination. Large banks of seeds may lie dormant in the soil for years awaiting the heat of fire to crack their bone-hard coats and allow the imprisoned embryos to germinate after winter rains. Most seeds from chaparral shrubs are difficult to germinate without special treatment to take the place of the heat of fire. Treatments include soaking seeds in acid or hot water, scarification (rubbing the seed coat away), or stratification (using cold temperatures to break or crack the seed coat).

Reproduction after fire is not the end to our story, however. While chaparral shrubs are busy resprouting or growing from seed, their place is temporarily occupied by other plants in a process called succession. The stages of succession in chaparral fall out something like this:

1. The first year, prolific annual wildflowers, bulbs, and grasses spring up by the thousands, mantling the burnt landscape with green after winter rains. Some of these plants have seeds that have lain dormant for more than twenty years.

   Bulbs are present under chaparral shrubs at other times as well, but because of shade and competition for water they send up only a few leaves each year. The ash from fires spurs on growth at an almost unbelievable rate, and the full light enhances growth as well, resulting in massive displays of flowers from the formerly flowerless bulbs. Bulb-bearing wildflowers include wild onions (*Allium* spp.), brodiaeas, globe tulips, mariposa tulips, and zygadenes.

   Several annual wildflowers actually appear only after fire, for their seeds require the heat of fire to crack their tough coats. Other annuals are stimulated to germinate in greater numbers after fire, but do not require it. Annuals that respond to fire include wind poppy (*Stylomecon heterophylla*), flame poppy (*Papaver californicum*), eucrypta, whispering bells (*Emmenanthe penduliflora*), and suncups (*Camissonia* spp.). The floral displays in the first year after a fire are often unsurpassed for many years running.

   At the same time that the original shrubs are resprouting, other opportunistic, fast-growing shrubs germinate from long-dormant seeds. Such shrubs include such members of the soft chaparral as California sagebrush, coyote bush, sticky monkeyflower, yerba santa, and black sage, as well as short-lived perennials and other normally rare shrubs (for example, golden eardrops, bush mallows, and deer broom lotus).

2. By the second year, the original chaparral shrubs that have resprouted together with new shrub seedlings have begun to fill in the holes that the fire created. There is less room for the light-loving annuals, grasses, and bulbs, which decline. The opportunistic shrubs are reaching full size and begin to bloom prolifically.

3. By the third and fourth years, annuals and bulbs have declined dramatically. Opportunistic shrubs continue to grow vigorously and bloom well, but the original chaparral shrubs are now regaining some of their former stature.

4. By the fifth, sixth, and seventh years, the originally dominant shrubs have nearly reached full size and are blooming well. They have shaded out many of the opportunists, which grow poorly in less than full light. In full cycle, from original shrubs that were killed outright, seedlings have now reached mature size.

Chapter 3

As mentioned before, woo[...]
and woodlands—with gaps b[...]
woodlands including riparian [...]
evergreen forest.

Ripa[...]

Riparian woodland is found only along permanent strea[...] and rivers, where the water table remains at or just below the surface all year. Our area has no true rivers, but there are several perennial streams—some with fairly broad floodplains—that support riparian woodland.

Because riparian woodlands have a guaranteed water supply, their component trees are very different from trees in most other environmental situations. These trees are not limited by the hot, dry days of summer; rather they can afford to grow fast and profligately right through the longest days of the year. Consequently, the derivation of riparian woodlands is entirely different from that of the rest of our flora. The closest relatives to riparian species come from summer-wet climates such as those across the Midwest and eastern parts of the United States. Wander in these forests if you are homesick for the look and feel of eastern hardwood forests.

Riparian trees—because of their ancestry from eastern United States climates—behave as though they still were adapted to cold winters. Nearly all are deciduous in winter, for they can afford to make whole new sets of leaves the following year, come what may. Leaves are also designed in ways that suggest water wastefulness: they're broad, thin, often lobed or compound, and held horizontally—fully exposed to the summer sun. They are also borne in thick tiers from top to bottom. Riparian trees reach maturity quickly because they're able to grow over such long periods each year. Quick growth may cause weak wood, however, and many riparian trees are liabilities because of their brittle limbs. They also have relatively short life spans.

Other features of riparian trees include wind pollination. High branches along stream corridors are wind tunnels, so wind becomes more reliable than insects for moving pollen long distances. By contrast, the shrubs and smaller plants along streams are mostly insect- or bird-pollinated.

Finally, riparian trees also use the wind for seed dispersal. Most have winged fruits (the samaras of maples and ashes and the winged seeds of alders) or hair-covered seeds (those of willows, cottonwoods, and sycamores). A few also use squirrels and other rodents, by offering edible nuts (the acorns of oaks, seeds of bay, and nuts of walnut and hazelnut).

Each riparian area has a different mixture of trees; tree species vary according to how hot summers are, how broad the floodplain is, and the accidental factor of which species arrive in a new area first. Because of the abundance of light (streams create corridors of light in otherwise heavily shaded forests) and water, there may be two or even three strata of trees of varying height. The tallest trees include bigleaf maple (*Acer macrophyllum*), box elder (*Acer negundo*), sycamore (*Platanus racemosa*), Fremont cottonwood (*Populus fremontii*), and white alder (*Alnus rhombifolia*). The understory always includes various willows (*Salix* spp.), sometimes in company with California hazelnut (*Corylus cornuta californica*) or elderberries (*Sambucus* spp.). Other woody plants may include California bay laurel (*Umbellularia californica*), California black walnut (*Juglans hindsii*), California buckeye (*Aesculus californica*), and—along the periphery—valley and coast live oaks (*Quercus lobata* and *agrifolia*).

...ce of water, abundance of
...Likewise, riparian woodlands
...rticularly common in Califor-
...adapted to climb other plants,
...need lots of water for fast
...f our vines are actually woody
...nds winding their way up and over
...erbaceous, without wood, and die
...year to thei...

Common vines include...r starlike creamy flowers in spring or early summer and great puffs of feathery-styled fruits in fall; native morning glories, with their arrowhead-shaped leaves and shallowly funnel-shaped white to pale pink flowers; vine honeysuckle, with its fuzzy, pale green paired elliptical leaves and modest racemes of pale pink, two-lipped flowers followed by red berries; and—perhaps most striking of all—California wild grape, with its bright green, palmately lobed leaves that turn fiery red and purple in fall, and which bears large masses of minute green flowers and sour purple fruits. All of these vines lend a tropical aspect to riparian woodlands in summer.

The lowest story consists mostly of small hedgelike growths of shrubs or large perennial plants. Among the hedge-formers are California wild rose, with pale pink-lilac petals and bright red-orange "hips" in fall; snowberry, with tiny pink bells hidden under leafy branches and spongy white berries in fall; creek dogwood, with an abundance of broadly ovate, paired leaves on red twigs and modest flat-topped clusters of four-pointed, white starlike flowers followed by small white or bluish fruits in fall; and squawberry, with interlaced, strongly scented branches that loop over, modest racemes of pale yellow flowers, and fuzzy red berries in fall.

Large perennial plants to look for include the fiery red, two-lipped flowers of scarlet monkeyflower; the scarlet flared trumpets of hummingbird fuchsia; the tall stems of stinging nettle (watch out for these!); and the hollow tubes of scouring rushes and horsetails (*Equisetum hymale* and *arvense*). The latter are ancient and primitive plants with jointed, ribbed green stems impregnated with silica, and with dark cones bearing myriad minuscule spores instead of seeds. They have survived nearly unchanged for three hundred million years.

Oak and Foothill Woodlands

Probably no more characteristic treelands occur in our part of California than oak woodlands, for they typify large areas of low foothills in both the Coast Ranges and Sierra Nevada. From a distance oak and foothill woodlands look similar: relatively dense groupings of trees. Where these same trees are widely spaced apart, they're referred to as savannah. In both, oaks grow as broad, rounded umbrellas; only on closer inspection is it obvious that they are of several different species. In what is often called foothill woodland, where the terrain is steeper or higher in elevation gray pines and California buckeyes often join ranks with the oaks.

In contrast to chaparral, oak and foothill woodlands most often occupy gently sloping terrain. When they occur on steep slopes, they do so on north- or east-facing hillsides, where the hot summer sun does not linger all day. Chaparral prefers sun-drenched south- and west-facing slopes.

Oaks have developed two equally effective water-conserving strategies for the hot, dry summers they must endure: live oaks bear long-lived, leathery, evergreen leaves that resist wilting and are covered with a waxy layer to prevent drying out; deciduous oaks—called robles—produce thin, deciduous leaves that are shed when water supplies dwindle but are renewed during the peak of the wet winter-spring rains. Both kinds grow side by side, blue oaks with interior live oaks, valley oaks with coast live oaks, and canyon live oaks with California black oaks. Some, such as valley and coast live oaks, prefer canyon bottoms with a higher summer water table; others, such as blue and interior live oaks, live perched high on hilltops or along the sides of rolling slopes.

Because of their sometimes prolific production of acorns—with their attendant rich stores of food—oaks are the intermediaries of their ecosystems, creating abundant food for insect larvae, rodents, and, not so long ago, Native Americans. Various midges and minute wasps also find oaks to their liking, as nurseries for their young; they lay eggs in various tissues of the oak and these grow into the multifarious galls we see commonly on oak trees. And even the parasitic mistletoes favor oak trees as places to grow and prosper; they bring with them birds that depend on their berries for food.

Woven into this web of life are the several shrubs and numerous grasses, wildflowers, and bulbs that benefit from the shaded protection of oak branches or the increased soil stability and water-holding properties of oak roots. In the more open oak woodlands, the wildflower displays can vie with those of our best grasslands.

Buckeyes and gray pines help fill out the personalities of these woodlands; while the rounded canopies of buckeyes mimic those of oak trees (but in miniature), the uneven and often double-barreled spires of gray pine punctuate and contrast with these umbrella shapes. Gray pine is the picture of a conifer well-adapted to dry, droughty summers: its sparse, gray needles reflect away summer sun and its stout trunks hold water reserves needed to complete the production of the oversized seed cones. Among the heaviest of all seed cones in the world, gray pine's also are armed with stout spine-tipped scales, but they offer up nutritious food. The large pine "nuts" are similar to those of the desert- and drought-adapted pinyon pines, and they are important to local animal life as yet another source of food. So too, doubtless, are the poison-laced chestnut-shaped seeds of the buckeye, for the poisons are not harmful to some animals. These seeds are also adapted for rolling, being perfectly round, and allow buckeyes to disperse their seeds downhill to the protection of shaded canyon bottoms.

The rich food reserve in the dominant trees—oaks, buckeyes, and gray pines—not only encourage animal dispersal of the seeds but give the seeds a head start when they germinate. Should they land in the shade of competing trees, the extra stored food allows the resulting saplings the chance to grow vigorously toward light.

While seed dispersal in oak and foothill woodlands therefore differs markedly from that in trees of riparian woodlands, pollination is another matter. All oaks and pines rely on wind to carry their pollen, just as with most riparian trees. Pollination occurs during late winter and early spring, just when winds are likely to be most reliable. The buckeye, however, uses another strategy. Its colorful candles of white flowers attract large numbers of pollinators, though the poisons in the nectar favor butterflies (which are immune to the poisons) over bees and other insects.

Mixed-evergreen Forest

Where oak canopies overlap, conditions favor a variety of other usually evergreen trees: California bay laurel, madrone, Douglas fir, tanbark oak (not a true oak but a *Lithocarpus*), and California nutmeg. The Douglas fir and California nutmeg are not met with in our own region, while the other trees are.

These mixed forests represent habitats intermediate in winter rainfall and summer drought between redwood forests, where summer fogs and heavy winter rains rule, and oak woodlands, where we've already seen the severity of summer drought. Often there will be no absolute line where oak woodland ends and mixed-evergreen forest begins, or where redwood forest holds sway and mixed-evergreen forest nudges the borders. Generally mixed-evergreen forests occur on north-facing slopes where south slopes are home to oak woodland. However, mixed-evergreen forests may carpet a canyon bottom alongside the narrow riparian corridor but give way to oak woodland or chaparral on adjacent slopes.

Often, too, the mixture of trees in these forests varies from locale to locale. Close to the coast, expect to see Douglas fir (*Pseudotsuga menziesii*) and California nutmeg (*Torreya californica*) in the forest; inland, expect to encounter canyon live oak, California black oak, California bay laurel, and madrone. The complex interactions of different trees from site to site are still not fully understood, for they also change with the age of the forest and its fire history.

With the exception of a few deciduous trees—California black and Garry oaks, California buckeye, and occasionally bigleaf maple—mixed-evergreen forests have the leathery, tough, evergreen leaves so characteristic of chaparral shrubs. Unlike those, however, mixed-evergreen forest tree leaves tend to be broader and—at least on lower branches—horizontally oriented, for purposes of more efficient light absorption for photosynthesis. Only near tree tops and only in some species (such as madrones) are leaves obliquely inclined, with pale undersides held skyward to reflect away intense summer sun.

Although in mixed-evergreen forests as elsewhere wind pollination is used for the conifers and oaks, both madrone and bay laurel differ sharply, having insect-pollination strategies. Madrone produces abundant, nectar-rich white bells in mid-spring (bee favorites); bay laurel makes long-lasting sets of small, pale yellow, saucer-shaped flowers from mid-winter to early spring. Bay laurel is thus especially important in sustaining insects active at a time of year when most life is dormant. It joins ranks with the manzanitas in fulfilling this important role.

As to seed dispersal, strategies resemble those of oak woodlands, again with many nutrient-rich stored foods in extra-large seeds. Only the madrone makes bright red-orange berries, attractive to large numbers of birds. Where Douglas fir occurs, its seeds are winged and wind distributed. This makes good sense, for Douglas fir is taller than the other trees, and winds easily reach its tall branches laden with seed cones.

Not only do the roots of these trees extend outward for great distances to pick up as much of the winter rains as possible, but the competing understory plants—shrubs, bunchgrasses, bulbs, and perennial herbs—seek water for later use. This intense competition for water means that the drier areas with least winter rainfall, where mixed-evergreen forest is marginal at best, have poorly developed understory vegetation. At the opposite pole, along the edge of redwood forests, the understory may be rich and varied. Most smaller plants are perennial; the annual life cycle is not favored by the relatively low light intensities. Many of these smaller plants extend into adjacent communities. The moisture-loving kinds extend into redwood forests and the droughty kinds—especially the few bunchgrasses, such as melicas and California fescue—into oak woodlands.

Redwood Forest

Redwood forests are represented in our area by isolated fragments found in the deepest, most protected canyon bottoms within reach of summer fogs. Although historically we know that other stands of redwoods once clothed coast-facing slopes in the Oakland and Berkeley hills, redwood forest was never a major plant community in this part of California.

Redwoods are remarkable trees that extend back in time to the beginnings of the cone-bearing trees called conifers. Once, great forests of various kinds of redwoods covered large tracts in North America, Europe, and Asia. Now they exist only in protected pockets as relicts from a time when the climate was more uniformly wet and had moderate year-round temperatures. The present distribution of coast redwood—canyon bottoms and slopes in the fog belt of coastal central and northern California—reminds us of how these trees are prevented from growing elsewhere. To the north, winters become too cold; to the south, summers are too hot and winters have too little rain; to the west, heavy salt-laden winds thwart growth next to the ocean; and to the east, summers lack fog and are too hot. Redwoods are also restricted from climbing higher than two- to three-thousand feet, owing to cold winter temperatures.

Redwoods—if allowed to grow unhindered for hundreds of years—exclude other trees by their tall, needle-covered branches that effectively shade out everything else. Where virgin redwood forest grows unimpeded, few smaller trees or shrubs live happily under the deep shade that their branches create. Yet redwood forest habitat is full of berry-producing shrubs. These favor forest edges and streamsides, especially in second-growth forests, where immature trees have not completed their overshadowing canopies.

In the deep shade of mature redwood forests live several smaller, herbaceous plants, such as various ferns, and sword fern (*Polystichum munitum*) in particular. Also in this shade are redwood sorrel (a ground-cover-forming oxalis); various violets; inside-out flower; various members of the lily family (such as trillium, false Solomon's seal, fetid adder's tongue, and bead lily); wild ginger; and several saxifrages (sugar scoops, fringe-cups, piggyback plant). All must do with short, periodic bursts of sunlight, and all take advantage of their locales by vegetative means of increasing their territory. Many of these forest denizens are limited in abundance in areas with minimal winter rainfall or only periodic summer fogs; to see the redwood forest understory at its best, journey to Humboldt and Del Norte counties in the northwestern extreme of our state.

Two fascinating aspects of redwood forest plants include the abundance of fleshy-fruited, berry-producing shrubs along streams, where birds depend on them for food and so help in their dispersal; and the many ant-dispersed seeds in the shade of mature redwoods. Ant-dispersed seeds have easily-seen white elaiosomes (oil bodies) appended to the main seed body. Ants are attracted by them, carry the seeds away, eat the elaiosomes, and discard the main seed with its embryo. Unrelated plants—trillium, fetid adder's tongue, western bleeding heart, smooth yellow violet, and inside-out flower—have hit upon this strategy as the best bet for moving their seeds.

Because deep shade creates cool, moist conditions most of the year, redwood-forest-floor plants have broad, water-wasteful leaves with maximum surface area to trap as much of the sun's light energy as possible. Trail plant (*Adenocaulon bicolor*), western coltsfoot (*Petasites palmatus*), and redwood sorrel all show thin, broad leaves that wilt easily in strong summer sun, yet that manage to remain turgid and healthy in the refreshing shade of forest aisles. Many plants here even have highly divided, fernlike leaves for efficient trapping of light energy: western bleeding heart, inside-out flower, and baneberry are examples.

Despite the fact that redwoods create a very special niche for low-growing herbaceous plants—cool, moist, acid soils—these plants are seldom exclusive to redwood forests. Many other coastal forests provide the same cool, moist conditions. So although closed-cone pine and Douglas fir forests, for example, are missing from our area, they are home to the same array of plants.

Fire and flooding have helped to maintain redwood forests where otherwise redwoods might be outcompeted by other kinds of trees. Redwood bark resists burning, since it lacks pitch and sap; mature trees also recover their fire wounds efficiently. Flooding may uproot old redwoods, but seeds are adapted to germinate in litter-free, sun-drenched soils such as those left behind after floods. And trees not uprooted by floods may send roots upward toward the surface or build a whole new set of roots near the surface even when the trunk and roots have been deeply buried under silt.

Redwoods are also efficient at replacing themselves when they're burned to the ground or, in the case of human intervention, at growing after being felled by logging. Dormant buds at the base of each tree are the secret; they grow into stump sprouts every year but are inhibited from growing more than a few feet tall by hormones produced by the top crown of the parent trunk. Once that source of hormones has been eliminated, the stump sprouts are free to grow, and grow they do. Circles of these sprouts become rings of mature trees in a relatively short time. This ability to regrow is what has saved many now-protected redwood forests that have been logged one or more times.

Notes

Section II
Cross-references, Keys, and Encyclopedias of Plants

How to Use a Key

The idea behind identifying plants with a key is simple. The key consists of a series of steps with choices. At each step you make one choice and proceed to the particular next step that choice indicates. When you arrive at a name, that's the plant. This book is organized so that there are separate keys to:

- Trees
- Vines
- Shrubs
- Wildflowers (herbaceous plants lacking wood or bark)

Each key is organized according to what seem the best criteria for simplicity:

- Trees according to whether or not they have colorful flowers (chances are you will find, on close observation, that your tree has tiny inconspicuous greenish or brownish flowers), then according to various leaf and flower characteristics.
- Vines according to flower color, then other obvious features.
- Shrubs according to flower color, then other features.
- Wildflowers according to form of flower, then flower color, then other features.

Wildflowers are the most difficult to key simply because they number far more species. Trees and vines are easiest because of far fewer choices.

Any step in the key compares similar characteristics; for example, if the step's first choice talks about flower color, the second choice does, too. Often two characteristics are used for each choice, so that if you're unsure of one characteristic, you can always use the second. For example, a step may ask you to compare one combination of flower color and number of stamens with another.

The greatest hurdle for the beginner is understanding the language used in keys. This is because, despite the attempt to simplify, there are certain technical terms and flower part names that must be used. If you're just beginning, turn to the glossary for diagrams and explanations of the various flower parts (pp. 309). Also, any time you run across a new word, such as "tendril," look it up in the glossary. Simple line drawings help clarify definitions in the glossary.

The other common hurdle is consistency between the key and the plant you're keying; an oddball plant may have six petals instead of the five the key indicates. The best policy when beginning is to look at more than one flower; if all flowers have six petals, then that probably is the number normally found.

A final hurdle is finding the stage of growth that the key asks for. For example, you may need to have flowers to identify your shrub, or fruits for your wildflower. Fruits and flowers don't always occur on the same plant at the same time, but these keys have been designed as much as possible for those that do have flowers and fruits simultaneously.

When you first start to key, remember these hints:

- Always start at the first step.
- Always consider all the choices at every step.
- Write down the numbers of the choices you follow; if you are uncertain at some particular

step, note this also. That way, if you arrive at a flower identity that doesn't fit the flower you're keying, you can go back to the steps you were uncertain about in the key.

- Be sure to look up words you don't know in the glossary.
- When possible, key while you're in the field. There you can use a plant in good condition and can see all parts of the plant.
- A good hand lens is indispensable for seeing the smaller flowers or parts of a flower in greater detail. A 10X is your best bet; it gives adequate magnification for most purposes. Hand lenses are available at nature stores or bookstores specializing in nature books.
- As soon as you've made an identification, check it in the encyclopedia section with the description and drawings given.

A Word about Naming

Both common names and scientific names are used throughout the chapters on trees, shrubs, vines, and wildflowers. Since common names are neither constant nor fixed, there is always disagreement about which to use. I have used the common names I feel are currently in greatest use, and coined names where no well-established ones existed before.

Scientific names may also change, for a variety of reasons. New research on relationships between species or genera, the discovery of earlier validly published names, and new points of view bring about such changes. Without a review of all the scientific literature the average naturalist is at a loss to know about these name changes and is seldom privy to the reasons for them.

A good case in point is *The Jepson Manual*, the newly published authoritative volume that supercedes *A California Flora* by Philip Munz. Since most recent books base their scientific names on Munz's work, and because many names have since been changed, the new manual will create confusion on naming for some time to come.

In this book, I have attempted to integrate the new naming as I feel best suits the situation. In many cases, I have retained the older, more familiar names, but then I have indicated the new names in brackets [ ]. Reasons for retaining older names are several: to retain names that are recognizable because they are familiar, to retain old names because I don't endorse the new treatments, and to avoid undue confusion. When the new names seem to elucidate relationships better, I have used them.

A Word about Plant Entries

In each of the following chapters—trees, shrubs, vines, and wildflowers—families, genera, and species are discussed in a consistent hierarchy. For each family cited key traits that members of the family share are described. Following that, genera and/or genus and species are described in some detail. Each of this second level of entries begins with the common name, then scientific name (in parentheses), then a page number to refer to in Appendix A. There you'll find information on where to go to see the species described in this book. Page numbers are not given for nonnative, alien, weedy, or introduced plants since their distribution often differs from year to year.

Notes

Chapter 4
Trees of the East Bay

Cross-references to Tree Names
Trees

TREE FAMILIES		
Scientific name	*Common name*	*Page number*
Aceraceae	Maple family	50
Betulaceae	Birch family	42
Caprifoliaceae	Honeysuckle family	48
Cupressaceae	Cypress family	44
Ericaceae	Heather family	46
Fagaceae	Beech (or oak) family	38
Hippocastanaceae	Horsechestnut family	48
Juglandaceae	Walnut family	60
Lauraceae	Laurel family	50
Myricaceae	Wax myrtle family	60
Myrtaceae	Myrtle family	52
Pinaceae	Pine family	54
Platanaceae	Plane tree family	56
Salicaceae	Willow family	62
Taxodiaceae	Redwood family	58

TREE GENERA AND SPECIES		
Common name	*Scientific name*	*Page number*
Alder, white	*Alnus rhombifolia*	42
Bay, or bay laurel California	*Umbellularia californica*	50
Bayberry, California	*Myrica californica*	60
Box elder	*Acer negundo*	52
Buckeye, California	*Aesculus californica*	48
Cottonwood, Fremont	*Populus fremontii*	62
Cypress, Monterey	*Cupressus macrocarpa*	44
Elder, box	*Acer negundo*	52
Elderberry, blue	*Sambucus mexicana*	48
red	*S. racemosa*	48
Gum, blue	*Eucalyptus globulus*	52
Hazelnut, California	*Corylus cornuta*	42
Juniper, California	*Juniperus californica*	44
Madrone	*Arbutus menziesii*	46
Maple, bigleaf	*Acer macrophyllum*	50

Key to Trees

1

- Seeds in cones; leaves needles or scales, go to 2
- Seeds inside seed pods, nuts, or berries; leaves usually broad (not scales or needles), go to 3

2

- Cones woody; leaves needles, go to 2A
- Cones woody; leaves scales, go to 2D
- Cones fleshy; leaves scales
  Juniperus californica (California juniper), p. 44

2a

- Cones remain on trees indefinitely, a foot or less long, go to 2B
- Cones fall when ripe, usually much more than a foot long, go to 2C
- Cones fall when ripe, cones less than 3 inches long
  Sequoia sempervirens (coast redwood), p. 58

2b

- Needles bright green, more than 6 inches long
  Pinus radiata (Monterey pine), p. 54
- Needles pale green, much less than 6 inches long
  Pinus attenuata (knobcone pine), p. 54

2c

- Needles droopy and sparse, gray
  Pinus sabiniana (digger or gray pine), p. 54
- Needles bushy and not drooping, gray-green
  Pinus coulteri (Coulter pine), p. 56

2d

- Scales deep green, seed cones more than 2 inches across
  Cupressus macrocarpa (Monterey cypress), p. 44
- Scales gray or dull green, seed cones less than 2 inches across
  Cupressus sargentii (Sargent cypress)

3

- Leaves lobed or compound, go to 4
- Leaves simple and unlobed, go to 12

4

- Leaves palmately compound
  Aesculus californica (California buckeye), p. 48
- Leaves pinnately compound, go to 5
- Leaves lobed, go to 8

5

- Leaves alternate
  Juglans hindsii (black walnut), p. 60
- Leaves opposite (the whole leaf, that is), go to 6

6

- Only 3 leaflets per leaf; fruit winged samaras
  Acer negundo (box elder), p. 52
- More than 5 leaflets per leaf; fruit a berry, go to 7

7

- Flowers in pyramid-shaped clusters; fruits red
  Sambucus racemosa (red elderberry), p. 48
- Flowers in flat-topped clusters; fruits bluish
  Sambucus mexicana (blue elderberry), p. 48

8
- Leaves palmately lobed, go to 9
- Leaves pinnately lobed, go to 10

9
- Leaves alternate; bark mottled
  Platanus racemosa (western sycamore), p. 56
- Leaves opposite; bark not mottled
  Acer macrophyllum (bigleaf maple), p. 50

10
- Leaves shallowly lobed, bluish green
  Quercus douglasii (blue oak), p. 42
- Leaves deeply lobed, green, go to 11
- Like the last but lobes have bristly tips
  Quercus kelloggii (California black oak), p. 42

11
- Branches droop nearly to ground; acorns long and slender
  Quercus lobata (valley oak), p. 40
- Branches not especially long and drooping; acorns wide and squat
  Quercus garryana (Garry oak), p. 40

12
- Leaf edges smooth, not toothed, go to 13
- Leaves toothed (look closely at several leaves), go to 17

13
- Leaves evergreen, tough and leathery, go to 14
- Leaves deciduous, thin, go to 16

14
- Leaves unscented, pale whitish underneath
  Quercus chrysolepis (canyon live or goldcup oak), p. 40
- Leaves strongly scented, not paler underneath, go to 15

15
- Leaves sickle shaped, camphor scented
  Eucalyptus globulus (blue gum), p. 52
- Leaves narrowly lance shaped, bay scented
  Umbellularia californica (California bay laurel), p. 50

16
- Leaves narrow and whitish gray
  Salix hindsiana (sandbar willow), p. 62
- Leaves lance shaped and pale or dull green
  Salix lasiolepis (arroyo willow), p. 62

17
- Leaf edges doubly serrate, go to 18
- Leaf edges with equal sized teeth only, go to 19

18
- Leaves soft and downy to the touch; small bushy trees
  Corylus cornuta californica (California hazelnut), p. 42
- Leaves not especially downy to the touch; tall trees
  Alnus rhombifolia (white alder), p. 42

19
- Leaves thin and deciduous; seeds hair-covered, go to 20
- Leaves leathery and evergreen; seeds lack hairs, go to 21

20
- Leaves broad and deltoid, teeth coarse
  Populus fremontii (Fremont cottonwood), p. 62
- Leaves narrow lance shape, teeth fine
  Salix laevigata (red willow), p. 62

21
- New bark smooth and orange-brown
  Arbutus menziesii (madrone), p. 46
- New bark greenish or dark brown, not especially smooth, go to 22

22

- Leaves narrowly oblong, fragrant on warm days

  Myrica californica (California bayberry), p. 60
- Leaves broadly elliptical to ovate, not fragrant, go to 23

23

- Leaves clear green on both sides, go to 24
- Leaves pale whitish green underneath

  Quercus chrysolepis (canyon live or goldcup oak), p. 40

24

- Leaf edges curl under; tiny clumps of hairs underneath at main vein junctions

  Quercus agrifolia (coast live oak), p. 40
- Leaf edges flat; no clumps of hairs underneath

  Quercus wislezenii (interior live oak), p. 40

Encyclopedia of Trees

BEECH FAMILY (FAGACEAE).

Woody shrubs or trees, leaves variable, sometimes lobed, sometimes deciduous. Flowers unisexual; male flowers borne in long, dangling chains and consisting mainly of stamens; female flowers produced in small clusters and consisting mainly of a cup of bracts surrounding a single pistil and 3-lobed feathery red stigma. Fruit a nut called an acorn.

Live or Evergreen Oaks (*Quercus* spp.). P. 296.

Oaks are one of the cornerstones of California's foothill country, dotted over rolling hills or forming nearly solid forests on steep, north-facing slopes. Our deciduous oaks look closely similar to species found in the eastern United States, but our live oaks have an entirely different aspect. All live oaks have tough, leathery undivided leaves, often with teeth along the edges. The young saplings in particular have these spiny teeth to protect against browsing. In addition, live oaks share with the deciduous kinds the rounded canopy, long dangling chains of male flowers (catkins), small clusters of inconspicuous female flowers—basically bract-enclosed ovaries topped by three feathery red stigmas—and in fall nutlike acorns that sit in a scale-covered cup. Nor do live oaks live exclusively apart from deciduous oaks. Often the two mix in canyon bottoms or along hill tops. The "live" way of doing things is one method of dealing with extreme drought, for it takes more energy and water to make a brand new set of leaves each year. Yet deciduous oaks may live in equally water-poor habitats; they simply lose their leaves early, and rely on sufficient winter rains to allow leaf renewal in spring.

All oaks also share in a variety of important ecological factors: all are infected by parasitic flowering plants called mistletoe (*Phoradendron villosum*); all are prone to a variety of galls caused by wasps laying their eggs on some part of the tree; and all produce acorns laced with bitter tannins that make them unpalatable to certain animals, edible to others. Mistletoes are particularly abundant on weakened trees, but often they occur sporadically even on healthy ones. In the evolutionary process mistletoe plants have lost their roots, and so penetrate living layers of bark to obtain the needed water and minerals that ordinary roots would otherwise provide. This is in spite of mistletoe leaves being green and able to photosynthesize. Mistletoes make minute, green, wind-pollinated flowers, and they bear heavy crops of clear white berries on which birds love to feed. As they feed, they attempt to wipe away the sticky seeds by rubbing their beaks against nearby branches. If there is a breach in the branch's natural defenses, the mistletoe seed sprouts a protuberance and starts feeding.

Galls are found in great variety on oaks, although they also occur on most other groups of plants. The gall itself is akin to a tumorlike growth, produced by the plant in response to its tissues being irritated by tiny wasps laying their eggs in there. These nonstinging wasps occur in wide variety—each kind preferring a particular group of plants and a particular part of the plant. As a consequence, galls are shaped according to the kind of wasp laying its eggs. The most conspicuous oak galls look like apples when fresh (green balls blushed red) and hang from oak branches. Other oak galls occur on leaves and come in varied forms (see blue oaks below). Inside each gall is a chamber for the eggs to hatch in; the larvae then eat and burrow their way out before flying off. Although galls only harbor "guests" once, they often remain on the tree long after their usefulness has ceased; those applelike galls turn dark brown and become spongy in age. Native Americans made use of them then for dyeing and as tinder for starting fires.

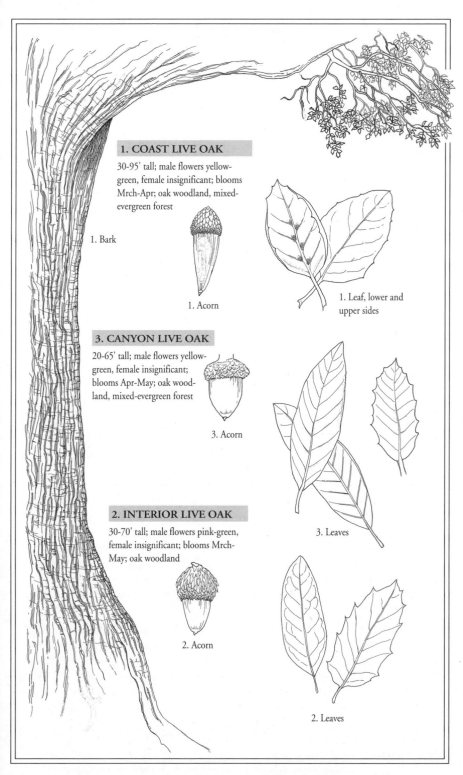

1. COAST LIVE OAK

30-95' tall; male flowers yellow-green, female insignificant; blooms Mrch-Apr; oak woodland, mixed-evergreen forest

1. Bark

1. Acorn

1. Leaf, lower and upper sides

3. CANYON LIVE OAK

20-65' tall; male flowers yellow-green, female insignificant; blooms Apr-May; oak woodland, mixed-evergreen forest

3. Acorn

3. Leaves

2. INTERIOR LIVE OAK

30-70' tall; male flowers pink-green, female insignificant; blooms Mrch-May; oak woodland

2. Acorn

2. Leaves

Acorns are produced in differing abundance according to the climate and other environmental conditions in a particular year—overburdening oak branches in good years, and barely appearing in poor. Usually different oak species find different years "good," so that most animals that rely on acorns for food find some in most years. A wide variety of animal life depends in part on these acorns, from tiny burrowing insects to large rodents. The latter often harvest acorns in fall, caching them for winter use and forgetting some in the process. This forgetfulness and caching behavior creates a ready means for oaks to distribute their acorns to new places; they even get planted at the same time. Native Americans made wide use of acorns as the staple in their daily diets—as mush or bread—but had to grind and leach them before they were palatable. Today, you will still find grinding rocks worn down where through long years acorns were pulverized by pestles. The acorn flour was then placed in a sandy area along the edge of a stream, and the water eventually leached away all bitterness. The process sometimes took days to complete.

We have three live oaks in our area: coast live oak (*Quercus agrifolia*), interior live oak (*Q. wislezenii*), and canyon live or goldcup oak (*Q. chrysolepis*). Although the first two are closely related, coast live oak favors areas closer to the coast. When it occurs inland, look for it in well-watered canyon bottoms (as around Mt. Diablo). Interior live oak is drought-adapted to hot, dry summers and frequents the hills inland. Also, coast live oak is an altogether bigger, beefier tree, and it is easily recognized by the curled-under leaf margins and tiny clumps of wooly hairs along main vein junctions underneath the leaf (use a hand lens). By contrast, interior live oak has flat leaves, sometimes toothed and sometimes not (coast live always bears sharp teeth), and it has no hairs underneath the leaf.

Canyon live oak differs by having lighter gray bark than either of the other two, and it has leaves with strikingly different colors above and below. Above, they're dark green; below, they're pale gray-green and, when young, sprinkled with fine gold powder. This gold powder is also typical of the fat acorn cups and gives rise to the common name goldcup oak as well as the species name "chrysolepis," which comes from the Greek for "golden scale."

Deciduous Oaks (*Quercus* spp.). P. 296.

Our deciduous oaks often remind newcomers to California of the oaks they knew "back east," for the leaves drop in autumn, the branches are bare through winter, and a brand new tapestry of foliage appears in spring. Further, the leaves are lobed much in the fashion of many eastern oaks. Our deciduous oaks have the other characteristics commented on under live oaks above. Like them, deciduous oaks are also vulnerable to mistletoe and various galls. Also like them, deciduous oaks have acorns that provide food for many different kinds of animals, and the acorns were used in the same ways by Native Americans.

We have three relatively common deciduous oaks. A fourth species, Garry oak (*Quercus garryana*), is only occasional. The most majestic of all is the great valley oak (*Q. lobata*), with a close resemblance to the eastern white oak (*Q. alba*). Valley oak seeks canyon bottoms with permanently high water tables, although it never grows with its roots directly by a stream or river. Sadly, valley oak was once the pride of the Central Valley, where valley oak woodlands predominated over vast expanses, but it is nearly extinct there due to agriculture and urbanization. Even those trees that have been preserved around housing developments slowly succumb to soil compaction and summer watering. The branches on old valley oaks droop so low they often nearly touch the ground, and fine old specimens may live more than two hundred years. Valley oak is identified by its whitish-gray, deeply checkered bark, deeply pinnately lobed leaves, and broad, rounded crown.

1. VALLEY OAK

40-110' tall; male flowers yellow-green, female insignificant; blooms Mrch-Apr; canyon bottoms in oak woodland

1. Acorn

2. Acorn

2. CALIFORNIA BLACK OAK

30-80' tall; male flowers sometimes reddish green, female insignificant; blooms Apr-May; oak woodland

3. BLUE OAK

20-65' tall; male flowers yellow-green, female insignificant; blooms Apr-May; oak woodland

3. Acorn

By comparison California black oak (*Q. kelloggii*) is a tree with narrower, more ascending crown and shallowly fissured dark gray to blackish bark; hence its inclusion in the black oak group. Although its leaves, too, are deeply pinnately lobed, each lobe tip ends in a sharp bristle. Leaves turn a lovely golden or yellow-brown color in fall. California black oak is at its best, however, in earliest spring, when the flush of new leaves wears a soft rosy haze from the numerous, velvety pink hairs covering them. Soon the hairs are shed; leaves quickly turn green. Look for California black oak on steep north-facing hillsides. They're particularly abundant in Morgan Territory, east of Mt. Diablo.

The third deciduous oak is blue oak (*Q. douglasii*), a generally smaller tree of rolling inland foothills, where it accompanies interior live oak and digger pine. Blue oak has received bad press. Some ranchers believe it to be a "weed tree," but actually the roots help stabilize soil and retain moisture rather than taking moisture away. True to its name, the fully developed leaves have a decidedly blue-green cast, and they are also distinctive because of their shallow lobes. Another distinctive blue oak feature is the long, fissured whitish gray bark (not checkered). Confusion sometimes arises, however, when blue oaks and valley oaks hybridize. Blue oaks are also noted for their wide variety of leaf galls; some resemble sea urchins or starfish, and others resemble shallow saucers or dishes.

BIRCH FAMILY (BETULACEAE).

Deciduous shrubs or trees from moist environments; leaves doubly serrated. Flowers tiny, unisexual, and at least the male borne in long, narrow catkins. Male flowers with floral bracts and stamens; female flowers with floral bracts and single pistil with superior ovary. Fruit a follicle with winged seeds or a nut.

White Alder (*Alnus rhombifolia*). P. 296.

Alders are one of our most reliable indicators of a permanently high water table; they favor the immediate edge of perennial streams and rivers. Even if the stream goes dry above ground, there will be ample water just a short way below. These graceful tall trees are leafless in winter; they regain their leaves just after the blossoms appear. Being wind pollinated, the male flowers dangle back and forth in long chains at winter's end. Just after, the shorter female catkins (which are plumper) push out their feathery stigmas. Although male and female catkins occur on the same tree, they mature at different times to avoid self-pollination. Later the female catkins grow into what pass for miniature green cones, which turn dark brown before they drop in autumn. Each female "cone" produces hundreds of tiny, winged seeds; these also depend on wind for long distance travel. White alder is a close relative to the more coastal red alder; the latter has leaf margins that curl under and the underside is sprinkled with tawny hairs; white alder is hairless, with flat leaves. The roots of both were used to extract a reddish dye.

California Hazelnut (*Corylus cornuta californica*). P. 296.

California hazelnut has leaves similar in shape and toothing to those of alders, but they're covered with soft, downy hairs. Also tied to moist environments, hazelnuts may or may not follow streams if they live in shaded forests; look for them in the hills overlooking the Bay, particularly on the edge of redwood forests. Hazelnut is a small tree with multiple trunks and horizontally tiered branches—again like alders, winter deciduous. It announces the end of winter by extending its long, very slender dangling male catkins. The female flowers—borne on the same plant—are harder to see, for they occur singly or in pairs from rounded buds. They consist of a bract-enfolded pistil with three feathery dark red stigmas (use a hand lens). As the

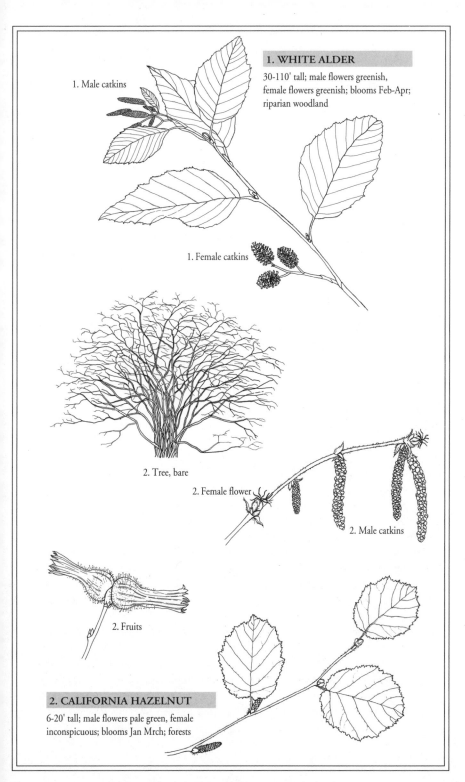

1. Male catkins

1. WHITE ALDER

30-110' tall; male flowers greenish, female flowers greenish; blooms Feb-Apr; riparian woodland

1. Female catkins

2. Tree, bare

2. Female flower

2. Male catkins

2. Fruits

2. CALIFORNIA HAZELNUT

6-20' tall; male flowers pale green, female inconspicuous; blooms Jan Mrch; forests

new leaves unfold, hazelnut is transformed into a pretty green treelet through summer. By summer's end, the delicious nuts have ripened and become enclosed in long beaked bracts covered with stiff nettlelike hairs. The taste of the nut is exactly like that of the cultivated filbert—delectable roasted—but the few nuts set on our trees are almost sure to be found first by hungry squirrels. Other nuts are lost to the predation of insects that lay eggs inside the developing nut. At fall's end, leaves turn a pale yellow before dropping, but preluding this event you'll see the buds for the long slender male catkins. They wait all winter long before opening.

CYPRESS FAMILY (CUPRESSACEAE).

Mostly trees with tiny scalelike leaves and shreddy bark. Tiny male pollen cones and larger female seed cones, which are either woody or fleshy.

Monterey Cypress (*Cupressus macrocarpa*).

Monterey cypress is rare in its true homeland—the Monterey Peninsula—where it grows into picturesquely tortured shapes on coastal cliffs. (It's included here since it is so well established in foothill California.) Monterey cypress has several attributes accounting for its wide popularity as a wind break, revegetation tree, or in large-scale gardens: rapid growth to maturity; high tolerance of wind and salt spray; and adaptation to sandy, nutrient-poor soils. Despite this, Monterey cypress is relatively short lived, and limbs become brittle with age, falling to disastrous effect during wind storms. All of our cypresses have thick, densely arranged scales; in Monterey cypress they're dark green. Other identifying features include horizontally trending branches, which give an irregular canopy, and globe-shaped woody seed cones with shield-shaped scales. Unless weather turns hot or there's been a fire, these cones fail to open despite age and, even when open, remain tightly connected to their branches. Such "serotinous" cones indicate long adaptation to a habitat where lightning-caused fires have brought about regeneration of the vegetation through reseeding.

California Juniper (*Juniperus californica*). P. 296.

Strictly speaking, California juniper is seldom a full-fledged tree, for it is low growing (seldom topping twenty feet) and usually has multiple trunks. Notwithstanding, it is a common component of the hottest inland foothill woodlands in California's interior and extending into the high desert mountains of southern California, where it joins the pinyon pine. On Mt. Diablo, its associates include digger pine, California buckeye, and blue oak. Although the mature plants have leaf scales similar to those of cypresses, the young trees—and the mature after injury—produce whorls of sharp, needlelike leaves. This is an adaptation to prevent browsing. Junipers differ from cypresses when they cone, for cypress cones are woody while juniper cones become fleshy and berrylike. The flesh is often purplish and covered with a powder, with few hard seeds inside. Instead of relying on fire to release the seeds, junipers entice birds to feast on their cones. The smell of juniper "berries" may seem familiar; in fact, a Caribbean species provides the flavoring for gin, although the word "gin" is derived from the Latin word for juniper. Another difference between junipers and cypresses is the dioecious condition of most junipers; male trees bear minuscule yellow pollen cones, and female trees bear seed cones.

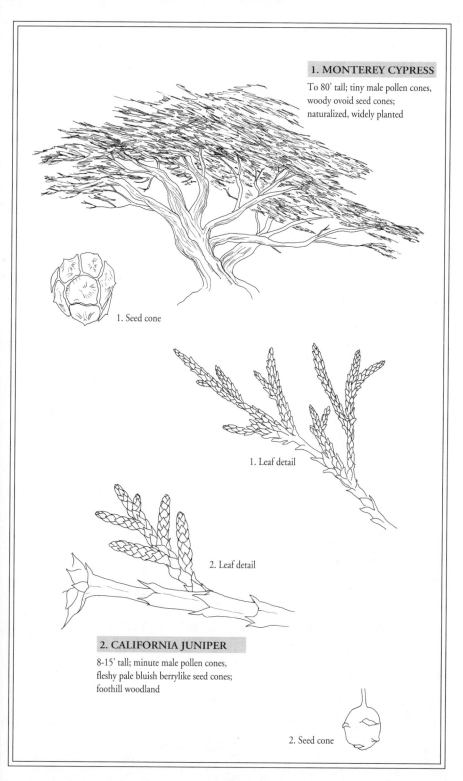

1. MONTEREY CYPRESS

To 80' tall; tiny male pollen cones,
woody ovoid seed cones;
naturalized, widely planted

1. Seed cone

1. Leaf detail

2. Leaf detail

2. CALIFORNIA JUNIPER

8-15' tall; minute male pollen cones,
fleshy pale bluish berrylike seed cones;
foothill woodland

2. Seed cone

Trees or shrubs with leathery, simple, often evergreen leaves. Flowers often bell- or urn-shaped; 5 sepals, 5 mostly fused petals, 5 separate stamens (not fused to petals) with tiny holes in the ends of anthers, and single pistil with usually superior ovary. Fruit a berry or capsule.

Madrone or Madroño (*Arbutus menziesii*). P. 296.

Madrone is often confused with its cousins, the manzanitas. But there are several apparent differences to the discerning observer: manzanita bark is generally smooth and a rich red-purple while madrone bark is smooth and red-brown when young but scaly later; manzanita leaves have smooth edges whereas madrone leaves are finely serrated; manzanita berries are smooth and apple-shaped but madrone berries are warty and strawberrylike. In fact, the latter reason is why a Mediterranean species of arbutus is called the strawberry tree. Its large, deep red-orange fruits are covered in bumps that from a distance look like the speckling on the outside of strawberries. In addition, madrone leaves lie horizontally rather than vertically as on manzanitas, and are dark green above. In summer, these same leaves may be tilted upward and inward to hold their backsides toward the sun, an attempt to reduce water loss since the underside is pale, almost silvery.

Madrones often form colonies when damaged, much in the manner of bays, but some truly huge old trees may also be single trunked. Then their identity is not so certain, for the old bark at the tree's base (and what is most apparent to the average naturalist) consists of fine brown scales. It's the new bark that is smooth and colorful. There are, however, some individuals in which the bark is at first green, then later orange-red. An herbal tea was formerly brewed from the older, scaly bark. When they burst into flower in mid-spring, madrones are among our most showy trees. Then the myriad compound clusters of white, urn-shaped, lily-of-the-valley-like blossoms are honey scented and attract hordes of honey and native bees. The red berries ripen in late fall, in time to tempt cedar waxwings and a number of other birds.

Look for madrones in mixed forests with other evergreen trees: frequent companions include coast live oak and California bay laurel and, closer to the coast, Douglas fir (*Pseudotsuga menziesii*) and tanbark oak (*Lithocarpus densiflorus*). Some of the most impressive trees occur, of all places, on rocky hill tops; there, the water table is evidently perched high.

Flower detail

Flowers

MADRONE

20-120' tall; flowers white;
blooms Mrch-May; forests,
canyon bottoms

Fruit

HONEYSUCKLE FAMILY (CAPRIFOLIACEAE).

Small trees or shrubs with opposite leaves that lack stipules. Flowers variable in size, sometimes regular, sometimes not; 5 minute sepals, 5 petals attached to a disc, 5 stamens, and a single pistil with inferior ovary. Fruit often a berry.

Elderberries (*Sambucus racemosa* and *mexicana*). P. 297.

Elderberries may start life as multitrunked shrubs, but they often mature into small trees, sometimes even with sizeable, oaklike trunks. Both kinds favor canyon bottoms where water is available in summer. However, red elderberry (*S. racemosa*) is mostly coastal—staying to the fog belt in our area—while blue elderberry (*S. mexicana*) is heat tolerant and somewhat less thirsty. Look for it throughout inland canyons. Both kinds have pairs of pinnately compound leaves with reputed medicinal qualities; crush a leaf if you've been stung by nettles and the pain is temporarily relieved. Pyramid-shaped clusters of creamy flowers crown branches of red elderberry in mid-spring, while flat-topped clusters of similar flowers announce spring's end for blue elderberry (they often flower with California buckeye). Rich nectar sources, the blossoms of both are soon replaced by berries: bright red for red elderberry (but poisonous) and pale blue for blue elderberry. The latter have been used in many ways by different peoples; Native Americans often dried them for later use, while pioneers made jam or wine from them. By late fall, elderberry branches stand leafless, but their arched newer branches, ribbed bark, and twigs with a soft central white pith easily identify them. Native Americans often trimmed back shrubs to obtain long, straight new branches, which were used for ceremonial clappers, gambling sticks, or flutes.

HORSECHESTNUT FAMILY (HIPPOCASTANACEAE).

Trees with opposite, palmately compound leaves and spikelike clusters of flowers; flowers slightly irregular; 5 sepals, 5 nearly separate petals, 5 stamens, single pistil with superior, single-chambered ovary. Fruit a leathery pod with one large, round, shiny brown seed.

California Buckeye or Horsechestnut (*Aesculus californica*). P. 296.

Horsechestnuts are not really chestnuts at all, even though the famous poem about the village smithie referred to horsechestnut and not true chestnut! The confusion comes about because the large, rounded seeds look so much like those of the edible chestnut (genus *Castanea*), which is a member of the beech family Fagaceae (and thus related to oaks). It would be a terrible mistake to attempt to use these seeds in place of chestnuts, for they're violently poisonous. Even Native Americans usually left them alone (except to stupefy fish in streams) except during years of poor acorn harvest; then the seeds were gathered and carefully detoxified by cooking, mashing, and leaching. The other common name—buckeye—alludes to the appearance of seed pods when they split in fall; the round seed peering through the covering of the pod looks like a large eye. California buckeye has adapted to life in the dry, hot foothill country, where it uses its ability to shed leaves in winter to lose them earlier if need be. (In especially dry years, buckeyes lose leaves in July.) Every season brings beautiful change to these umbrella-canopied trees: in winter, the white-gray bark shines after rains and displays its patina of lichens; in spring, the delicate new leaves unfurl to apple green; in late spring or early summer, branches wear candles of white flowers; in early fall, pear-shaped seed pods decorate branches as leaves fall. If you examine a flower cluster carefully, you'll find only a few flowers with a pistil; this is to prevent too many heavy seed pods from growing on any one branch, weighing too heavily.

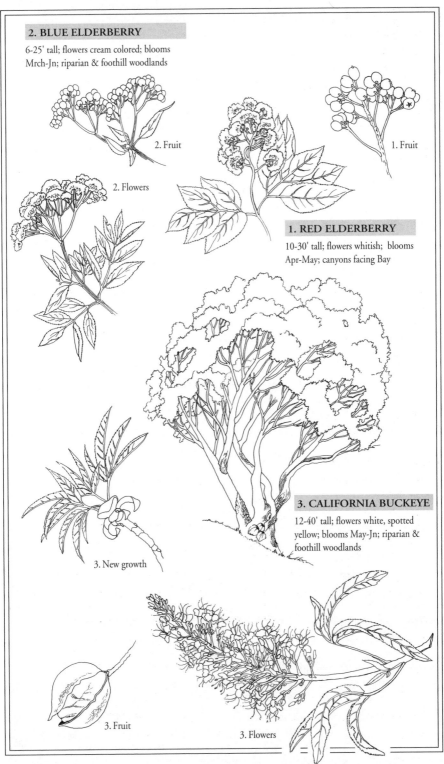

2. BLUE ELDERBERRY

6-25' tall; flowers cream colored; blooms
Mrch-Jn; riparian & foothill woodlands

2. Fruit

1. Fruit

2. Flowers

1. RED ELDERBERRY

10-30' tall; flowers whitish; blooms
Apr-May; canyons facing Bay

3. CALIFORNIA BUCKEYE

12-40' tall; flowers white, spotted
yellow; blooms May-Jn; riparian &
foothill woodlands

3. New growth

3. Fruit

3. Flowers

LAUREL FAMILY (LAURACEAE).

Trees with leathery, evergreen, fragrant, simple leaves. Flowers generally small; 2 or more series of tepals, and several series of stamens in 3s, the stamens with overarching "hoods" and single pistil with superior ovary. Fruit a fleshy drupe.

California Bay Laurel (*Umbellularia californica*). P. 297.

California bay laurel is one of our most widespread and versatile native trees. It occurs in the yellow pine belt of the Sierra or at the edge of coastal Douglas fir and redwood forests, or it is mixed inland with other evergreen trees, such as live oaks and madrone. In the latter situations, it seeks canyon bottoms with high water tables, as it does most often throughout the foothills of southern California. California bay laurel is one of those trees with multiple personalities, owing to its wide adaptability and variability. In southern Oregon (where it reaches its northern limits), it's known as Oregon myrtlewood; in the redwood country of northern California many call it pepperwood; and in our area it's most often simply called bay or bay laurel. A true member of the mostly tropical laurel family, California bay is closely related to the bay tree of the Mediterranean (*Laurus nobilis*), which has served since time immemorial as a tree of power and knowledge, and whose leaves are most often used in cooking authentic Mediterranean dishes. California bay's leaves may substitute, but be warned that their aroma and flavor is much stronger; moderation is needed. The strong oils that give bay its distinctive smell also are responsible for driving away weevils and other insect pests. Indians used them to protect acorn granaries, but on hot days they may also give you a headache. Despite this odor, deer closely browse young bay trees in times of drought; evidently the newer leaves are not so fibrous or malodorous.

Bays are malleable in terms of their overall shape. Near the coast they assume a variety of shapes according to the wind patterns, and on flood plains they grow into venerable old trees with huge trunks; inland they grow in more shapely fashion, and most become multitrunked. The handsome, lance-shaped, shiny leaves are on the branches year-round, but the small clusters of yellow flowers enliven the scene in winter and early spring and are one of the few nectar sources then for bees. Flaplike valves over the stamens protect the pollen from getting wet during this rainy time of year. By summer, small green fruits are beginning to enlarge. By late fall, they've reached full size, turning purple before being plucked by birds or falling for rodents to find. The young fruits look like miniature avocadoes, even to the pattern of dappled lighter flecks on the green exterior and the large stonelike seed inside. These fruits offer a valuable clue to bay's relatives; avocadoes too are tropical members of the laurel family, that are native to Central American rainforests.

MAPLE FAMILY (ACERACEAE).

Shrubs or trees with opposite, often palmately lobed, deciduous leaves. Flowers borne in short to long chains and individually small; 5 sepals, 5 petals (sometimes missing), 10 stamens, and double pistil with superior ovary. Fruit a winged samara.

Bigleaf Maple (*Acer macrophyllum*). P. 296.

Bigleaf maple is truly big, both for its leaves, which may measure several inches across, and for the size of its canopy. (Our local specimens don't reach their full potential due to lack of summer water.) In our area, bigleaf maple seeks out canyon bottoms near permanent streams, but it truly reaches its best development in the heavy winter rains and summer fogs of the northwestern corner of our state. Bigleaf maple is particularly beautiful in mid-spring, when the

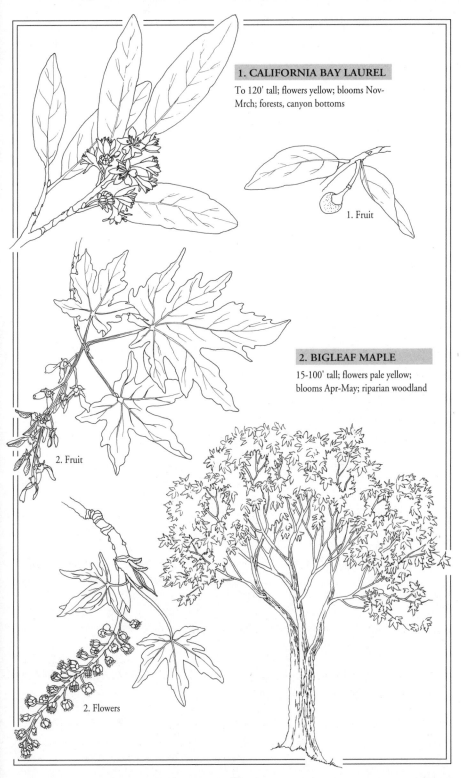

1. CALIFORNIA BAY LAUREL

To 120' tall; flowers yellow; blooms Nov-Mrch; forests, canyon bottoms

1. Fruit

2. BIGLEAF MAPLE

15-100' tall; flowers pale yellow; blooms Apr-May; riparian woodland

2. Fruit

2. Flowers

swollen winter buds burst open to release hanging chains of pretty pale yellow flowers and coppery new leaves. Bees are fond of bigleaf maple flowers; the results are the chains of double-winged fruits in fall. As fruits drop, each half severs, and the two parts go spinning on their way buoyed by winds and breezes. With cold fall nights, leaves turn a golden yellow before falling. Look for bigleaf maple as companion to alders, California bay laurel, and sometimes western sycamore, with its similar-looking leaves.

Box Elder (*Acer negundo californicum*). P. 296.

Unlike its sister species in the maple family, box elder has compound leaves, the leaves divided into three (sometimes as many as five) separate, prettily serrated and lobed leaflets. Other striking differences include the fact that box elder is wind pollinated and, as a result, is dioecious: male trees are draped with slender, pendant pinkish male flowers in spring, while female trees have shorter catkins of greenish pistillate flowers. In fall, however, there's no mistaking this as another maple, for the doubly winged samaras are shed in exactly the same manner as described for bigleaf maple. Box elder does not color up the way bigleaf maple does. Neither is it abundant in our area, reaching its best development only along broad flood plains of major streams and rivers, where its tall crowns can effectively compete with western sycamore, red alder, and cottonwoods.

MYRTLE FAMILY (MYRTACEAE).

Shrubs or trees with simple, usually highly aromatic, evergreen leaves. Four or 5 partly fused sepals, 4 to 5 usually-separate petals (sometimes missing), numerous stamens attached around a nectar disc, single pistil with inferior ovary. Fruit a berry or hard, woody capsule.

Blue Gum (*Eucalyptus globulus*).

The eucalyptuses are so familiar that many people assume they're California natives. Nothing could be further from the truth: they're from southeastern Australia and Tasmania and were introduced back in the late 1800s as a crop for a fast-growing source of lumber. Of the more than 600 species native down under the blue gum was chosen, though with little real merit for the purposes at hand; blue gum wood is exceedingly hard to work unless absolutely fresh. Meanwhile, thousands of trees had been planted across California's lowlands, where they persist in good health today, despite the idea that recent hard freezes would kill them; they merely resprout from undamaged roots. Blue gum eucalyptus is a unique tree in our land: the bluish juvenile leaves are paired, broad and horizontally oriented, while the adult leaves are narrow, scythe shaped, and vertically oriented. (These characteristics all represent adaptations to reduce water loss for high branches in full sun.) The heavy camphor odor in the leaves is widely believed to inhibit other plants, but the truth is that blue gums' roots are efficient at finding water and hogging it all for themselves. Certainly in their native homeland, myriad smaller plants grow beneath them without problem.

In flower, too, blue gums stand out: a warty cap falls off as the blossom opens. This cap represents the sepals and is the reason for the genus name, which comes from two Greek words: "eu" for well and "calyptus" for cap or covering. Instead of colorful petals, long masses of white stamens attract pollinators. In fruit, a second cap covers the woody bell-like seed pods.

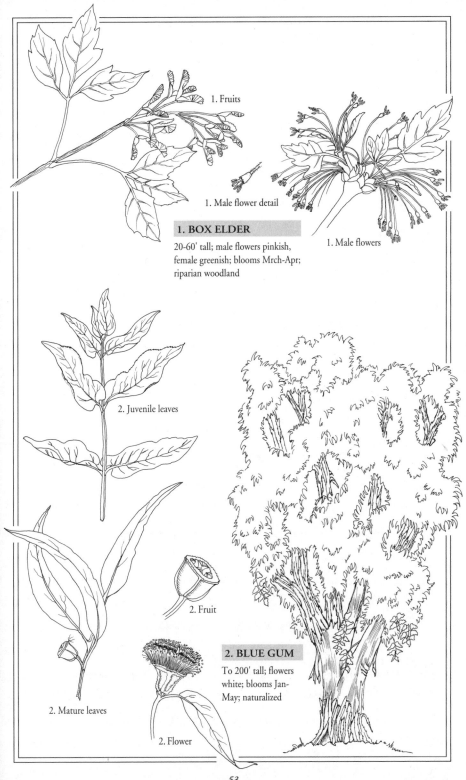

1. Fruits

1. Male flower detail

1. Male flowers

1. BOX ELDER

20-60' tall; male flowers pinkish,
female greenish; blooms Mrch-Apr;
riparian woodland

2. Juvenile leaves

2. Fruit

2. BLUE GUM

To 200' tall; flowers
white; blooms Jan-
May; naturalized

2. Mature leaves

2. Flower

PINE FAMILY (PINACEAE).

Trees with evergreen, usually fragrant needlelike leaves. Dense clusters of small, cylindrical pollen cones in spring, and smaller clusters of single female seed cones, with spirally arranged woody scales and usually winged seeds.

Closed-cone Pines (*Pinus attenuata* and *radiata*). P. 296.

The pines of our area fall into two natural groupings: closed-cone pines and big-cone pines. The first group is characterized by tough, strongly attached cones that remain on the tree its entire life. Even after the two years needed to produce mature seeds, cones remain tightly closed except in the event of unusually hot days or a fire. This adaptation to natural conflagrations, commented on above for Monterey cypress, is a way of rapidly reforesting a burned area. Most closed-cone pines are not long lived; they are genetically programmed to be renewed every thirty to forty years. Consequently, in preparation for fire, they bear heavy loads of cones, and they grow to maturity rapidly. For this reason, the Monterey pine has been widely planted as our most common pine for city streets, hillsides, and Bay Area gardens. The short life span becomes an unfortunate liability, for as trees age their limbs become brittle and fall readily during strong winds. Other liabilities include shallow, thirsty roots that absorb most of the nearby water (with the result of little undergrowth, as shown in the extensive pine forests in the Oakland and Berkeley Hills), and flammable resins in the needles. Monterey pines contributed to the explosive effects of the disastrous 1991 fire in Oakland and Berkeley.

Monterey pines occur naturally in only three major areas along California's central coast: Año Nuevo in northern Santa Cruz County, the Monterey Peninsula, and around Cambria on the southern Monterey coast near Hearst Castle. Yet despite their restrictions in the wild—owing to climate changes and poor competitive abilities—they are not only widely planted throughout lowland California but are the major timber tree in New Zealand, and very popular in Australia as well. Down under, Monterey pine is called by its species name—radiata, derived from the arrangment of seed cones in whorls around branches. Other distinguishing features include the long, thick, dark green needles in clusters of three and the warty appearance of the scales on the asymmetrical seed cones.

Our other closed-cone pine—knobcone pine (*P. attenuata*)—is adapted to hotter, drier climates, and it occurs naturally over a large but scattered range from central California north to the Oregon border. Our best local stand occurs on Knobcone Ridge on Mt. Diablo. Knobcone pines not only survive where summers are hot and dry but do it on nutrient-poor, rocky soils. Comely in youth, the trees become narrow and rangy in age, usually because so many seedlings compete for dominance after fire. The needles are in threes but are much shorter than those of Monterey pines, and the asymmetrical seed cones are distinctive by their stoutly spined scales.

Bigcone Pines (*Pinus coulteri* and *sabiniana*). P. 296.

Bigcone pines not only have truly large seed cones—up to a foot long and almost as broad—but are the heaviest of all the world's pine cones, with a weight of several pounds. Unlike the closed-cone pines, their mature cones fall to the ground after they ripen. By that time rodents and other would-be seed harvesters have already begun trying to get at the large, nutritious seeds. The seeds, in fact, which taste much like those of the pinyon pines, were widely used by Native Americans here for food. Adapted to droughty conditions, both species of bigcone pines grow well inland or on steep slopes where water retention is poor. Of the two, digger or gray pine (*Pinus sabiniana*) is the tougher, and it is our only pine that is well adapted to California's

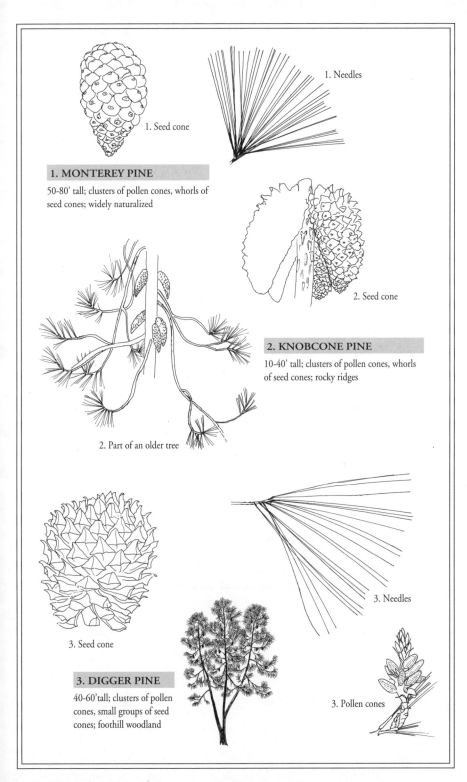

1. Seed cone

1. Needles

1. MONTEREY PINE

50-80' tall; clusters of pollen cones, whorls of
seed cones; widely naturalized

2. Seed cone

2. Part of an older tree

2. KNOBCONE PINE

10-40' tall; clusters of pollen cones, whorls
of seed cones; rocky ridges

3. Seed cone

3. Needles

3. DIGGER PINE

40-60' tall; clusters of pollen
cones, small groups of seed
cones; foothill woodland

3. Pollen cones

scorching hot summers and scant winter rainfall. The sparsely clustered, drooping grayish needles that accommodate these conditions result in a see-through appearance. Another distinctive feature is the multiple trunks, which may diverge near the trees' base (or sometimes higher). The enormous cones look like spiky pineapples, each stout scale ending in an out-turned spine. Beware standing under these trees when cones are ready to fall!

The Coulter pine (*P. coulteri*) is similar to the digger or gray pine, but the grayish green needles are thick and bushy (not droopy and sparse), and the trunk seldom splits. The seed cones are equally large but are slightly longer than broad (digger has cones about as long as broad at the base), and the spine-tipped scales turn up at their tips. Coulter pine reaches its northern limits around Black Diamond Mines; there are also good stands in Mitchell and Back canyons on Mt. Diablo. Typically digger pine follows other drought-worthy trees such as interior live oak, blue oak, and California buckeye. Coulter pines are more typical of middle elevations in coastal mountains where they join canyon live oak, coast live oak, California bay laurel, and madrone.

PLANE TREE FAMILY (PLATANACEAE).

Trees with deciduous, palmately lobed leaves and conspicuous stipules. Trees monoecious, with greenish, wind-pollinated flowers borne in dense ball-shaped clusters. Male with long stamens; female with single pistils. Fruits arranged in a spiny ball, which breaks apart when ripe.

Western Sycamore (*Platanus racemosa*). P. 296.

The word sycamore goes back a long way to the Old World, where it was first applied to a fig relative with sycamorelike leaves. Later it was applied to certain European maples as well as the Old World plane tree—a true sycamore—but only in the last few hundred years has it been used for our native sycamores, also of the plane tree family. Obvious resemblances occur between maples (genus *Acer*) and our sycamore, especially as concerns the thin, deciduous, strikingly palmately lobed leaves. Both also favor well-watered canyon bottoms and are dominant elements in our riparian woodlands. There the resemblances stop, for sycamore has beautifully patchworked bark, with different layers' colors exposing different ages, a jigsaw puzzle of cream, tan, and gray. The leaves are arranged alternately rather than oppositely, with conspicuous, collarlike stipules at their bases (these may fall away later as leaves age), and the leaf stalk is hollow at its base, fitting neatly over the pointed axillary bud. Axillary buds on most plants lie just above the junction of the leaf stalk with the stem. Sycamore flowers are also different, for they're strictly wind pollinated, and they occur in ball-shaped clusters borne on long, hanging racemes. The male balls are smaller, the female larger, becoming warty in fruit. Wind plays a role a second time for seed dispersal, since the substantial-looking female fruits shatter in late fall and the hair-lined seeds scatter on strong winds.

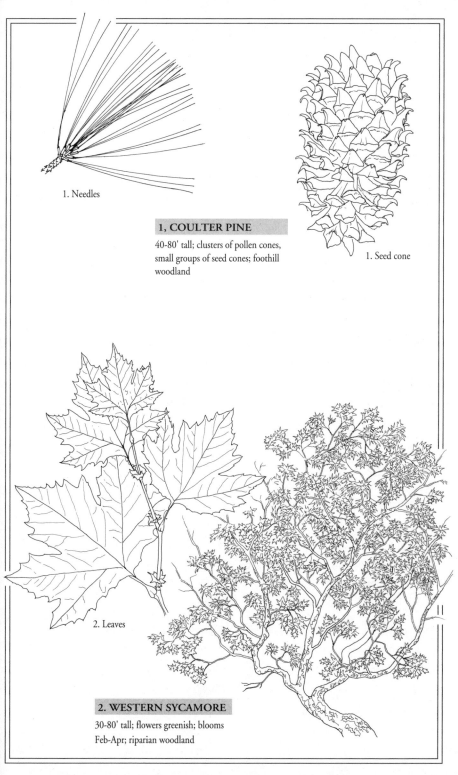

1. Needles

1, COULTER PINE

40-80' tall; clusters of pollen cones, small groups of seed cones; foothill woodland

1. Seed cone

2. Leaves

2. WESTERN SYCAMORE

30-80' tall; flowers greenish; blooms Feb-Apr; riparian woodland

Tall trees with fibrous, nonpitchy red-brown bark and needlelike leaves, which fall attached to twigs. Tiny yellow male pollen cones; woody female seed cones with whorled, wrinkled, diamond-shaped scales.

Coast Redwood (*Sequoia sempervirens*). P. 297.

Perhaps coast redwood is California's most famous native tree. Certainly it holds the record as the world's tallest (up to 367 feet) and is also counted among those with large girth (up to 18 feet in diameter) and long life (living to well over 1,000 years). From the fossil record we know that coast redwood once dominated forests across North America and Europe, yet today it has been isolated in a long strip along California's coast defined by climate: needed are foggy summers (for extra moisture and cool temperatures) and mild winters (without appreciable freezing or snow). Factors that limit growth include hot, dry summers, cold winters, and heavy winds and salt spray. Redwoods are notoriously shallow-rooted, so they topple in floods or heavy wind storms. Because of their thirsty, shallow roots, redwoods should never be planted near dwellings where roots may penetrate and clog water and sewage lines. In our area, redwoods are rather restricted—occurring mostly in canyon bottoms or on steep north-facing slopes, where they aren't exposed to hot, dry air in summer. Several stands in the Oakland hills have been logged, but in those same hills we see redwoods in Redwood Regional Park and around the hamlet called Canyon near Moraga. Since redwoods are on the edge of their range here, the forest understory is not nearly so lush as in parts of Sonoma or Mendocino counties, but you can still gain some idea of the feel of a forest on a foggy day, when water droplets condense on branches and drip to the ground or, in winter, when the dense canopy of branches filters out most of the light. Because of the gloom created by redwood branches and the persistent feeling of dampness, in California, Native Americans did not make their homes under redwood forests—a good practice for us all.

Redwoods are well known for the economic value of their wood. Thanks to the high tannin content in the bark and wood, dry rot (caused by fungi) is prevented, giving the wood its long life. Redwood is especially used for fences, shingles, and decking, where this antirot factor is particularly important. Were it not for a second feature—a circle of buds at the base of trunks capable of growing into new shoots—the felling of acre upon acre of redwoods would have resulted in the near total destruction of these forests. Happily, however, these stump sprouts grow into new trees around the parent tree, so that in second- or third-growth forests, the trees form rings.

Coast redwood is surprisingly adapted in two special ways: for fires and for flooding. While floods often uproot trees by undermining their roots, the silt deposited prepares a new seed bed for redwood saplings. If not too deeply buried, redwoods have a second strategy: they send their roots toward the surface or grow a whole new set of roots. In fire, healthy redwoods resist burning with their fibrous nonvolatile bark, but when they are badly burned, the ash from the fire also stimulates seedling germination and growth.

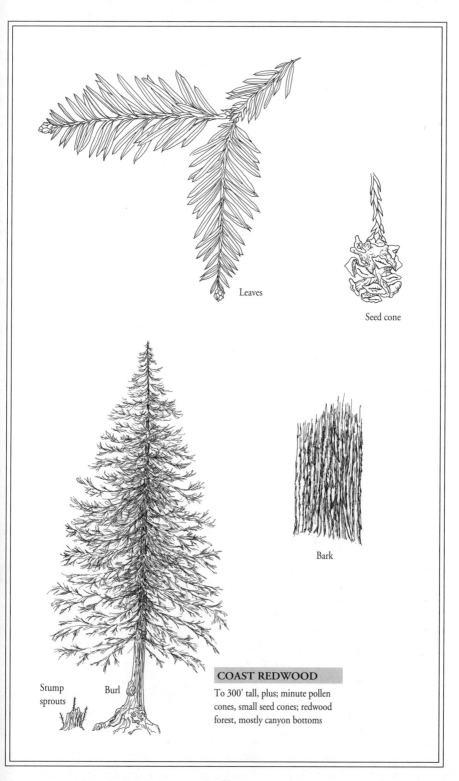

Leaves

Seed cone

Bark

Stump sprouts

Burl

COAST REDWOOD

To 300' tall, plus; minute pollen cones, small seed cones; redwood forest, mostly canyon bottoms

WALNUT FAMILY (JUGLANDACEAE).

Trees with deciduous, aromatic, pinnately compound leaves. Monoecious with wind-pollinated flowers, the male in long dangling catkins, the female in small clusters nearby. Male flowers a series of floral bracts and long stamens; female flowers a single pistil enclosed in bracts with 2 feathery stigmas protruding. Fruit a fleshy drupe with nutlike stone inside.

Black Walnut (*Juglans hindsii* [*J. californica hindsii*]). P. 296.

California has two native walnuts, but it's this species of black walnut that has played the greatest role for the commercial walnut crop; all English walnuts are grafted onto the vigorous black walnut rootstocks. It's not uncommon to see untended walnut orchards reverting to shoots from their rootstock, for hidden buds below the graft often sprout more vigorous growth than the scion grafted on. Walnut Creek is probably one of the original native homes for black walnut. This walnut is a relict from a time when California's climate was generally wetter year-round. As an indication of this, natural groves of wild walnuts occur strictly as riparian trees, often mixed with alders, maples, willows, and cottonwoods. Because Native Americans were fond of walnuts, they spread the nuts into canyons where they weren't originally found in the north Coast Ranges and Sierra foothills, so it's hard to be sure of their origin where we see them today. Although black walnut "meats"—actually rich, oil-storing cotyledons of the seed—are highly flavorful, the shells are so difficult to remove that black walnut is not a viable crop. Instead the hulls, which waft the same strong scent as the leaves on warm days, are useful as a dark brown dye.

WAX MYRTLE FAMILY (MYRICACEAE).

Shrubs or small trees with simple, aromatic, often leathery leaves. Flowers borne in dense short spikes between leaves near branch tips, and wind pollinated. Trees monoecious. Fruits corrugated, waxy purple berries.

California Bayberry or Wax Myrtle (*Myrica californica*). P. 296.

California bayberry reaches its eastern limits in the Bay Area and is found here only because of the moderating influence of summer fogs on the Bay's climate. Normally, bayberry is to be looked for immediately along the north coast, where it grows happily on the edge of pine or redwood forests. Here, by contrast, it occurs in a few protected areas in the Berkeley hills, as for example in Tilden Regional Park. A fast-growing small tree, bayberry is handsome all year, with glossy, lance-elliptical evergreen leaves that have finely serrated margins. The latter may be hard to see, for leaves often curl slightly along their edges and also may be wavy. On cold winter days the leaves don't seem to have any obvious scent, but on warm days—especially with the new growth—leaves are delightfully and subtly scented, not at all like their sound-alike but distant relative the bay laurel. The common names allude to properties of the small, dark purple, wrinkled fruits, for these are aromatic when crushed (a bit reminiscent of the true bay) and have a waxy substance in them (hence the name wax myrtle). Our western species has so little wax, however, that it would require a major effort to collect enough wax to make candles; the famed bayberry candles come from the waxier fruits of an eastern species. Nonetheless, that pleasant scent is not entirely lost, for the dried leaves add subtle flavor to stews and sauces.

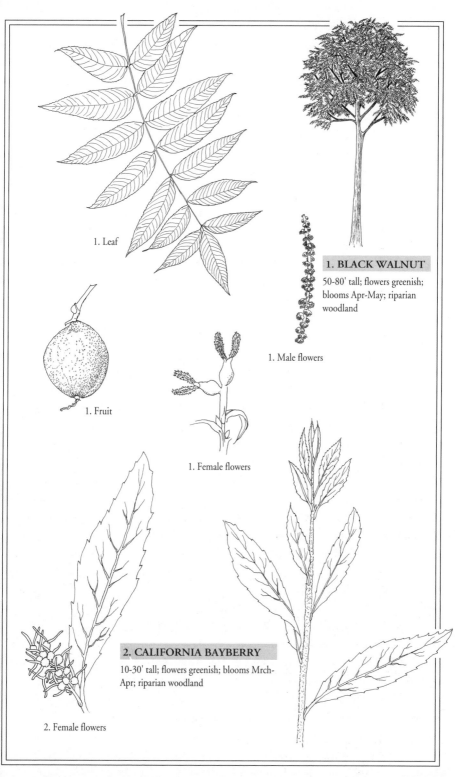

1. Leaf

1. BLACK WALNUT
50-80' tall; flowers greenish;
blooms Apr-May; riparian
woodland

1. Male flowers

1. Fruit

1. Female flowers

2. CALIFORNIA BAYBERRY
10-30' tall; flowers greenish; blooms Mrch-
Apr; riparian woodland

2. Female flowers

Deciduous shrubs and trees of riparian habitats, with simple leaves and deciduous stipules. Trees dioecious and wind pollinated. Male flowers borne in catkins, consisting of floral bracts and bunches of stamens; female flowers borne in similar-shaped catkins, consisting of floral bracts and single pistils. Fruit a capsule containing numerous, hair-covered seeds.

Fremont Cottonwood (*Populus fremontii*). P. 296.

One of the tallest and most dramatic riparian trees is Fremont cottonwood; even its range is impressive, for it lines streams throughout the Coast Ranges, Sierra foothills, eastside valleys, and even permanent desert oases. Relatively uncommon in our area, Fremont cottonwood prefers only the broadest flood plains along permanent streams or rivers. Growing to mature size more rapidly than any other trees save their relatives the willows, cottonwoods are well named for the masses of cottony hairs attached to seeds. During dispersal, it's not uncommon to see great masses of these seeds covering the ground like so much snow. This, in fact, is the only conspicuous phase of reproduction, for the small headlike flower clusters go nearly unnoticed. The genus name, "populus," alludes to an ancient Latin word for people, which in turn goes back to Greek. From these roots stem numerous words, including "population" in English and "pueblo" in Spanish. The name is fitting, for populuses do form populations wherever they grow: the parent tree sends out roots that soon send up suckers, which in their turn grow into more trees. Often a stand of cottonwoods will be from one original clone. The only other lowland cottonwood in California is the black cottonwood (*P. trichocarpa* [*P. balsamifera trichocarpa*]), not found in our area but scattered elsewhere. Its leaves are narrower and darker and lack the conspicuous teeth found on Fremont cottonwood leaves.

Willows (*Salix* spp.). P. 296.

Willows are our most ubiquitous small tree or large shrub accompanying permanent waterways. There, they sometimes dominate along narrow canyon bottoms; on broad flood plains they form the understory to taller trees such as maples, cottonwoods, and alders. Among our fastest growing plants, willows are very easy to start by simply burying bundles of twigs. Their stems contain a compound that encourages rooting in other plants. Since willow roots and grows with ease, it has great potential for containing steep slopes where rainfall is plentiful. Willows also lend seasonal interest to any landscape: through winter their twigs are often vividly colored in shades of brown, orange, or yellow; the young catkin buds are covered with silky white hairs (which is why they're called pussy willows). Pick these buds before they open at winter's end, but don't put them in water; dried, they retain their form indefinitely. Male trees are usually in the majority and wear masses of yellow, pollen-bearing catkins (hay fever sufferers beware); female trees have green catkins, the color of the ovaries. Later the white fluff unfolding from seed pods makes them a liability in a garden, for the fluff invades every corner and crevice.

Willows have long been noted for other special abilities: the pliable new twigs served admirably as the framework for baskets or for building temporary huts; and the bark is one of the original sources of salicylic acid, the active ingredient in aspirin. In fact the word salicylic is derived from the Latin genus name *Salix*, and one Spanish place name in the Bay Area from the same root is Sausalito ("sauce" means willow).

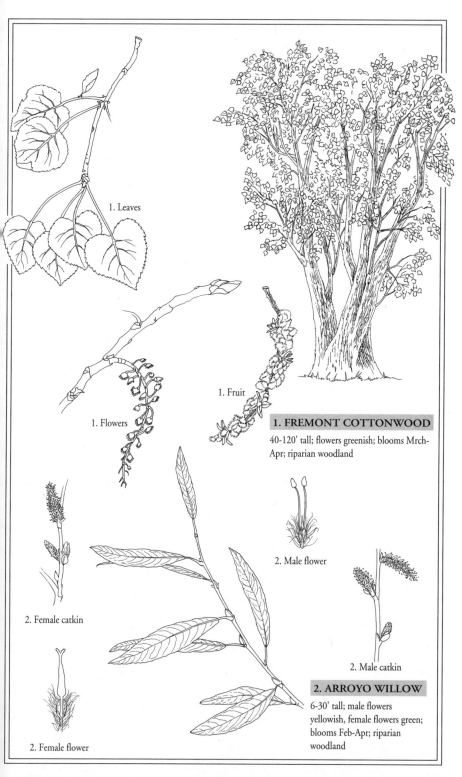

1. Leaves

1. Flowers

1. Fruit

1. FREMONT COTTONWOOD

40-120' tall; flowers greenish; blooms Mrch-Apr; riparian woodland

2. Male flower

2. Female catkin

2. Male catkin

2. ARROYO WILLOW

6-30' tall; male flowers yellowish, female flowers green; blooms Feb-Apr; riparian woodland

2. Female flower

Chapter 5
Shrubs of the East Bay

— Cross-references to Shrub Names begins on page 73 —

Photo Gallery
East Bay plants and plant communities

Oak woodlands wear a new crown of leaves in early spring. [photo Tim Dallas]

Plant communities form mosaics according to slope and soils. Backside of Mt. Diablo from Morgan Territory. [photo Glenn Keator]

Oak woodlands are dominated by several oaks: here coast live oak, with its muscular trunks. [photo Glenn Keator]

Interior live oak, with male catkins, favors hot, inner foothills. [photo Glenn Keator]

Oaks also house wide variety of galls: tumorlike growths that result from eggs layed by tiny wasps. [photo Glenn Keator]

Mistletoe is a densely branched greenish parasite dependent upon oaks. [photo Glenn Keator]

The rounded crowns of California buckeye also occur here. [photo Glenn Keator]

Digger pines join oaks on steep rocky slopes. Note their wispy needles and huge seed cones. [photo Glenn Keator]

Blue oak has deciduous, blue-tinted leaves. [photo Glenn Keator]

Hound's tongue pushes up new leaves in early spring. [photo Glenn Keator]

California buttercups light up the floor of oak woodlands in early spring. [photo Glenn Keator]

Look for the curious pipe-shaped flowers of California pipevine in early spring. [photo Glenn Keator]

Buckeyes show drama in their leafless fall and winter condition. [photo Glenn Keator]

In mid-spring, there's Ithuriel's spear. [photo Tim Dallas]

Mule's ear appears a bit later, around May. [photo Tim Dallas]

Mid-spring is also the time for Indian paintbrush and Chinese houses. [photo Tim Dallas]

Spring's end is announced by the clear pink petals and red throats of ruby-throated clarkia. [photo Glenn Keator]

Above, white alder grows rapidly, with a dense summer canopy. Canyon bottoms, with permanent streams, harbor the exuberant growth of riparian woodlands. [photo Glenn Keator]

Blue elderberry is laden with flat-topped clusters of creamy flowers brimming with nectar. [photo Glenn Keator]

Western sycamore is identified by its puzzlelike pattern of multi-colored bark. [photo Glenn Keator]

Box elder's winged fruits are designed for wind travel. [photo Glenn Keator]

Bigleaf maple's leaves are large and efficient at trapping light. [photo Glenn Keator]

Arroyo willows often line stream sides as a lower story. [photo Glenn Keator]

Scarlet monkeyflower favors openings along streams. Its curious flowers are designed for hummingbird pollination. [photo Glenn Keator]

Wild grape leaves festoon over shrubs and trees, with vivid colors in fall. [photo Glenn Keator]

Manroot climbs riparian canopies by its coiled tendrils. [photo Glenn Keator]

Clematis is a vine whose spectacular hairy white fruits are designed for wind transport. [photo Tim Dallas]

California wild rose creates prickly hedges with large lavender-pink flowers. [photo Glenn Keator]

Chaparral blankets hot rocky hillsides with a velvetlike cover of droughty shrubs. [photo Tim Dallas]

Jim brush smothers its branches in blue-purple blossoms in mid-spring. [photo Tim Dallas]

Manzanita produces beautiful clusters of urn-shaped flowers in late winter. [photo Glenn Keator]

Chamise is adapted to summer heat by its dense clusters of needlelike leaves. [photo Glenn Keator]

Bush poppy, with myriad yellow blossoms, favors the chaparral edge. [photo Glenn Keator]

Also on the edge is wooly-leafed paintbrush, whose roots hook onto those of chamise. [photo Glenn Keator]

Poison oak occurs in many places and in many guises: its tiny flowers are perfumed to attract bees. [photo Glenn Keator]

Fine displays of wildflowers, such as these brodiaeas and woodland stars, follow chaparral burns.
[photo Glenn Keator]

Many burned shrubs, such as toyon, regrow from long-dormant buds. [photo Glenn Keator]

— continued on page 133 —

Cross-references to Shrub Names

SHRUB FAMILIES

Scientific name	Common name	Page number
Anacardiaceae	Sumac family	112
Asteraceae	Daisy or composite family	86
Berberidaceae	Barberry family	82
Caprifoliaceae	Honeysuckle family	98
Cornaceae	Dogwood family	90
Ericaceae	Heather family	96
Fabaceae	Pea family	102
Fagaceae	Beech or oak family	82
Garryaceae	Garrya family	92
Grossulariaceae	Gooseberry family	94
Hydrophyllaceae	Waterleaf family	114
Lamiaceae	Mint family	100
Malvaceae	Mallow family	98
Oleaceae	Olive family	102
Papaveraceae	Poppy family	104
Rhamnaceae	Buckthorn family	84
Rosaceae	Rose family	106
Rutaceae	Rue or citrus family	112
Scrophulariaceae	Figwort or snapdragon family	90
Solanaceae	Nightshade or potato family	100
Thymeleaceae	Daphne family	90

SHRUB GENERA AND SPECIES

Common name	Scientific name	Page number
Ash, flowering	*Fraxinus dipetala*	102
Barberry, California	*Berberis pinnata*	82
Berry, coffee	*Rhamnus californica*	86
huckle-	*Vaccinium ovatum*	96
oso	*Oemleria cerasiformis*	108
service	*Amelanchier pallida*	106
sour-	*Rhus trilobata*	112
thimble-	*Rubus parviflorus*	110

Common name	Scientific name	Page number
Blue witch	*Solanum umbelliferum*	102
Brickel bush, California	*Brickellia californica*	88
Brooms	*Cytisus, Genista* spp.	102
Buckbrush	*Ceanothus cuneatus*	84
Buckthorn, redberry	*Rhamnus ilicifolia*	86
Bush sunflowers	*Ericameria* spp.	88
Chamise	*Adenostoma fasciculatum*	106
Chaparral pea	*Pickeringia montana*	104
Cherry, bitter	*Prunus emarginata*	110
choke-	*P. virginiana demissa*	110
hollyleaf	*P. ilicifolia*	
wild	*P.* spp.	110
Chinquapin, shrub	*Chrysolepis chrysophylla minor*	84
Coffee berry	*Rhamnus californica*	86
Coyote bush	*Baccharis pilularis*	88
Creambush	*Holodiscus discolor*	108
Currant, chaparral	*Ribes malvaceum*	94
golden	*R. aureum*	94
pink flowering	*R. sanguineum glutinosum*	94
Dogwood, creek	*Cornus* spp.	90
stream	*C.* spp.	90
Golden fleece	*Ericameria arborescens*	88
Gooseberry, canyon	*Ribes menziesii*	94
fuchsia-flowered	*R. speciosum*	94
hillside	*R. californicum*	94
oak-leaf	*R. quercetorum*	94
Greasewood	*Adenostoma fasciculatum*	106
Heather, mock	*Ericameria ericoides*	88
Holly, California	*Heteromeles arbutifolia*	106
Honeysuckle, twinberry	*Lonicera involucrata*	98
Hopbush	*Ptelea crenulata*	112
Huckleberry, evergreen	*Vaccinium ovatum*	96
Jimbrush	*Ceanothus sorediatus*	84
Leatherwood, western	*Dirca occidentalis*	90
Lilac, wild	*Ceanothus* spp.	84
Lupine, blue bush	*Lupinus albifrons*	104
Mahogany, mountain	*Cercocarpus betuloides*	106
Mallow, bush	*Malacothamnus* spp.	98
Manzanitas	*Arctostaphylos* spp.	96
Monkeyflower, bush	*Mimulus aurantiacus*	90
sticky	*M. aurantiacus*	90

Common name	Scientific name	Page number
Mulefat	*Baccharis salicifolia*	88
Nightshade, blue	*Solanum umbelliferum*	102
Ninebark	*Physocarpus capitatus*	108
Oak, leather	*Quercus durata*	82
poison	*Toxicodendron diversilobum*	112
scrub	*Quercus berberidifolia*	82
Ocean spray	*Holodiscus discolor*	108
Oso berry	*Oemleria cerasiformis*	108
Pea, chaparral	*Pickeringia montana*	104
Penstemon, gaping	*Keckiella breviflora*	92
red rock	*K. corymbosa*	92
Pitcher sage	*Lepechinia calycina*	100
Plum, Sierra	*Prunus subcordata*	110
wild	*P.* spp.	110
Poison oak	*Toxicodendron diversilobum*	112
Poppy, bush	*Dendromecon rigida*	104
Rose, wild	*Rosa* spp.	110
wood	*R. gymnocarpa*	110
Sage, black	*Salvia mellifera*	100
pitcher	*Lepechinia calycina*	100
Sagebrush, California	*Artemisia californica*	86
Silk tassel bush	*Garrya* spp.	92
Snowberry	*Symphoricarpos albus*	98
Sourberry	*Rhus trilobata*	112
Squawbush	*Rhus trilobata*	112
Thimbleberry	*Rubus parviflorus*	110
Tobacco, tree	*Nicotiana glauca*	100
Toyon	*Heteromeles arbutifolia*	106
White-thorn	*Ceanothus leucodermis*	84
Yerba santa	*Eriodictyon californicum*	114

Key to Shrubs

Flowers orange

1

- Flowers irregular, 2-lipped; leaves unlobed, lance shaped
  Diplacus aurantiacus (sticky monkey-flower), p. 90
- Flowers nearly regular, tubular; leaves unlobed, ovate
  Lonicera involucrata (twinberry honeysuckle), p. 98
- Flowers regular, bowl shaped; leaves often palmately lobed
  Fremontodendron californicum (fremontia or flannel bush)

Flowers red or maroon

1

- Flowers bright scarlet red, go to 2
- Flowers maroon and white, hanging bells, go to 4
- Flowers maroon, go to 5

2

- Flowers trumpet or fuchsia shaped, go to 3
- Flowers two-lipped
  Keckiella corymbosa (red rock penste-mon), p. 92

3

- Short subshrub with flared trumpetlike flowers
  Zauschneria californica (California fuchsia), p. 208
- Definite shrubs (woody throughout) with tubular fuchsialike flowers
  Ribes speciosum (fuchsia flowered gooseberry)

4

- New growth with spines between nodes as well as at nodes
  Ribes menziesii (canyon gooseberry), p. 94
- New growth with spines at nodes only
  Ribes californicum (hillside gooseberry), p. 94

5

- Flowers waterlilylike, with many tepals; scented like old wine
  Calycanthus occidentalis (western spicebush)
- Flowers like tiny 4-pointed stars; no discernible odor
  Euonymus occidentalis (western burning bush)

Flowers pink or rose

1

- Flowers irregular, pealike
  Pickeringia montana (chaparral pea), p. 104
- Flowers regular, not pealike, go to 2

2

- Flowers bell shaped, nodding or hanging, go to 3
- Flowers other shapes and not hanging, go to 8

3

- Leaves lobed, go to 4
- Leaves entire; not sticky, go to 6

4

- Stems armed with spines
  Ribes amarum (bitter gooseberry)
- Stems spineless, go to 5

5

- Flowers pink to rose; plants of forested areas
  Ribes sanguineum glutinosum (pink flowering currant), p. 94
- Flowers pale purplish-pink; plants of chaparral
  Ribes malvaceum (chaparral currant), p. 94

6

- Leaves stiff, smooth, without teeth
  Arctostaphylos spp. (manzanitas), p. 96
- Leaves stiff, smooth, shiny, finely toothed
  Vaccinium ovatum (evergreen huckleberry), p. 96
- Leaves thin, hairy, go to 7

7

- Plants erect with many branchlets
  Symphoricarpos albus (snowberry), p. 98
- Plants creeping
  Symphoricarpos mollis (creeping snowberry), p. 98

8

- Leaves pinnately compound; stems spiny, go to 9
- Leaves not compound; stems spineless, go to 10

9

- Flowers more than 2 inches across; hips retaining sepals
  Rosa californica (California wild rose), p. 110
- Flowers much less than 2 inches across; hips sepalless
  Rosa gymnocarpa (woodrose), p. 110

10

- Petals flat and flowers wide open; stamens unfused
  Prunus subcordata (Sierra plum), p. 110

- Petals forming a shallowly flared trumpet; stamens unfused; flowers very fragrant
  Rhododendron occidentale (western azalea)
- Petals forming a cup-shaped flower; stamens fused together to form a tube
  Malacothamnus spp. (bush mallow), p. 98

Flowers blue or purple

1

- Flowers distinctly pea like, with banner, wings, and keel
  Lupinus albifrons (blue bush lupine), p. 104
- Flowers not at all pea like, go to 1a

1a

- Many flowers densely clustered, each flower less than ½ inch across, go to 2
- Individual flowers larger, go to 3

2

- Flowers with 5 petals and sepals; leaves simple
  Ceanothus spp. (wild lilacs), p. 84
- Flowers with single petal; leaves pinnately compound
  Amorpha californica (false indigo)

3

- Flowers irregular, 2-lipped, go to 4
- Flowers regular, not 2-lipped, go to 6

4

- Flowers in circular whorls up stem
  Salvia mellifera (black sage), p. 100
- Flowers arranged otherwise, go to 5

5

- Leaves quilted, highly fragrant
  Lepechinia calycina (pitcher sage), p. 100
- Leaves not quilted, not fragrant
  Penstemon heterophyllus (foothill penstemon), p. 214

6

- Petals not joined
  Malacothamnus (bush mallow), p. 98
- Petals joined, go to 7

7

- Flowers flat or saucer shaped
  Solanum umbelliferum (blue witch),
  p. 102
- Flowers tubular with flared lobes
  Eriodictyon californicum (yerba santa),
  p. 114

Flowers yellow

1

- Leaves compound, go to 2
- Leaves simple (may be deeply lobed),
  go to 6

2

- Several leaflets per leaf, go to 3
- Three leaflets per leaf; no spines, go to 4

3

- Leaves pinnately compound; leaflets with
  spiny teeth
  Berberis spp. (barberry), p. 82
- Leaves palmately compound; no
  spiny teeth
  Lupinus arboreus (yellow tree lupine)

4

- Flowers not pea shaped, in spikes
  Rhus trilobata (squawbush), p. 112
- Flowers pea shaped in dense umbels,
  go to 5

5

- Branches mostly low sprawling
  Lotus scoparius (deer broom lotus), p.
  252
- Branches mostly standing upright
  Cytisus spp. (brooms), p. 102

6

- Flowers in daisylike heads, go to 7

- Flowers not daisylike, go to 11

7

- Leaves linear, sticky warm days, go to 8
- Leaves deeply lobed, not sticky
  Eriophyllum confertiflorum (golden
  yarrow), p. 196
- Leaves scalelike on broomlike branches, not
  sticky
  Lepidospartum squamatum (scale broom)

8

- Flower heads with small rays, individually
  very small
  Gutierrezia californica (California
  matchstick)
- Flower heads rayless, individually small,
  go to 9
- Flower heads with showy rays, individually
  larger, go to 10

9

- Floral bracts around head not in
  regular rows
  Ericameria arborescens (golden fleece),
  p. 88
- Floral bracts around head in well-
  defined rows
  Chrysothamnus nauseosus (rabbitbrush)

10

- Floral bracts around head in one even row
  Senecio douglasii (Douglas's bush senecio)
- Floral bracts around head in two or more
  rows, overlapping
  Ericameria linearifolia (bush sunflower),
  p. 88

11

- Stems armed with spines
  Ribes quercetorum (oak-leaf gooseberry),
  p. 94
- Stems not spiny, go to 12

12

- Flowers with tubes, go to 13

- Flowers lack tubes, go to 15

13
- Petals irregular (2-lipped); leaves opposite
  Lonicera subspicata (southern honey-suckle)
- Petals regular; leaves alternate, go to 14

14
- Leaves bluish green, unlobed
  Nicotiana glauca (tree tobacco), p. 100
- Leaves bright green, palmately lobed
  Ribes aureum (golden currant)

15
- Flowers in dense ball-like clusters; six tepals
  Eriogonum umbellatum (sulfur buck-wheat), p. 184
- Flowers not in ball-like clusters; four or five petals or tepals, go to 16

16
- Flowers pealike with banner, wings, and keel, go to 17
- Flowers small, 5-pointed stars, go to 18
- Flowers hanging bells
  Dirca occidentalis (western leatherwood), p. 90
- Flowers large, shallow saucers, go to 19

17
- Branches not spiny
  Cytisus spp. (brooms), p. 102
- Branches with daggerlike spines
  Ulex europaea (gorse)

18
- Leaves more than 2 inches long, elliptical
  Rhamnus californica (coffee berry), p. 86
- Leaves less than 2 inches long, nearly round
  Rhamnus ilicifolia (redberry buckthorn), p. 86

19
- Stamens numerous; petals clear yellow

Dendromecon rigida (bush poppy), p. 104
- Stamens, 5; petals golden yellow to yellow orange
  Fremontodendron californicum (fremontia)

Flowers greenish or brownish

1
- Leaves opposite, go to 2
- Leaves alternate, go to 3

2
- Flowers borne in long, dangling chains
  Garrya spp. (silk tassel bush), p. 92
- Flowers borne in small, upright bunches
  Forestiera neomexicana (desert olive)

3
- Leaves compound or deeply divided, go to 4
- Leaves simple, go to 5

4
- Leaves divided into 3 shiny green leaflets
  Toxicodendron diversilobum (poison oak), p. 112
- Leaves slashed into narrow, gray divisions
  Artemisia californica (California sagebrush), p. 86

5
- Leaves pale to gray green, aromatic
  Brickellia californica (California brickel bush), p. 88
- Leaves deep to bright green, not aromatic, go to 6

6
- Male flowers in long dangling chains, go to 7
- Flowers not in dangling chains, go to 8

7
- Leaves green, without hairs on upper surface

Quercus berberidifolia (scrub oaks), p. 82
- Leaves dusty gray green, with numerous starlike hairs on upper surface
Quercus durata (leather oak), p. 82

8
- Leaves fuzzy and dull
Cercocarpus betuloides (mountain mahogany), p. 106
- Leaves smooth and shiny, go to 9

9
- Leaves more than two inches long, elliptical
Rhamnus californica (coffee berry), p. 86
- Leaves much less than two inches long, rounded
Rhamnus ilicifolia (redberry buckthorn), p. 86

Flowers white

1
- Leaves opposite, go to 2
- Leaves alternate, go to 9

2
- Leaves strongly sage scented, go to 3
- Leaves not sage scented, go to 4

3
- Flowers less than 1 inch long, in whorls
Salvia mellifera (black sage), p. 100
- Flowers more than 1 inch long, in racemes
Lepechinia calycina (pitcher sage), p. 100

4
- Flowers irregular, 2-lipped, go to 5
- Flowers regular, go to 6

5
- Gaping flowers; lower lip with purple lines
Keckiella breviflora (gaping penstemon), p. 92
- Flowers not gaping; petals without purple lines

Lonicera subspicata (no common name)

6
- Leaves compound
Fraxinus dipetala (flowering ash), p.102
- Leaves simple, go to 6a

6a
- Leaves wedge shaped; ovary superior
Ceanothus cuneatus (buckbrush), p. 84
- Leaves ovate; ovary inferior, go to 7

7
- Twigs reddish; leaf margins smooth
Cornus spp. (creek dogwood), p. 90
- Twigs brownish; leaf margins smooth (and flowers in tight balls)
Cephalanthus occidentalis (western buttonbush or button willow)
- Twigs brownish; leaf margins smooth, lobed, or toothed, go to 8

8
- Flowers bell shaped, hidden beneath leaves
Symphoricarpos spp. (snowberry), p. 98
- Flowers not bell shaped, obvious beyond leaves
Viburnum ellipticum (native viburnum)

9
- Leaves compound or deeply divided, go to 10
- Leaves simple, go to 11

10
- Stamens numerous; fruit a fleshy "berry"
Rubus (blackberry and thimbleberry), p. 110, 126
- Stamens numerous; fruit of several inflated dry pods
Physocarpus capitatus (ninebark), p. 108
- Stamens, five; fruit single-seeded surrounded by a wing
Ptelea crenulata (hopbush), p. 112

11
- Leaves needlelike, clustered in bunches

Adenostoma fasciculatum (chamise), p. 106
- Leaves broad, not clustered in bunches, go to 12

12
- Bark red-purple; flowers bell or urn shaped
  Arctostaphylos (manzanitas), p. 96
- Bark dark brown; leaves broadly ovate, entire; flowers bell or urn shaped
  Gaultheria shallon (salal)
- Bark dark brown; leaves narrowly ovate, finely toothed; flowers bell or urn shaped
  Vaccinium ovatum (evergreen huckleberry), p. 96
- Bark green, gray, or brown; flowers not bell shaped, go to 13

13
- Petals joined to form a tube, go to 14
- Petals not joined; leaves not shiny on top, go to 15

14
- Flowers flared trumpet shape, very fragrant
  Rhododendron occidentale (western azalea)
- Flowers long tubular, no odor
  Eriodictyon californicum (yerba santa), p. 114

15
- Many flowers clustered together in heads; leaves round elliptical
  Baccharis pilularis (coyote bush), p. 88
- Like the last but leaves lance shaped and willow like
  Baccharis salcifolia (mulefat), p. 88
- Flowers not clustered together in heads, go to 16

16
- Leaves tough and evergreen, go to 17
- Leaves thin and deciduous, go to 19

17
- Leaves golden underneath; flowers in long slender spikes

Chrysolepis chrysophylla minor (shrub chinquapin), p. 84
- Leaves not golden underneath; flowers in other arrangements, go to 17a

17a
- Leaf margins smooth, mature bark whitish
  Ceanothus leucodermis (white-thorn), p. 84
- Leaves finely serrated, mature bark dark, go to 18

18
- Leaf surface shiny; flowers in racemes
  Prunus ilicifolia (hollyleaf cherry)
- Leaf surface dull; flowers in complex cymes
  Heteromeles arbutifolia (toyon), p. 114

19
- Leaves with fruity odor when crushed
  Holodiscus discolor (creambush), p. 108
- Leaves not particularly aromatic, go to 20

20
- Bark with circular rings of lenticels; astringent smell, go to 21
- Bark lacks circular rings of lenticels; not astringent smelling, go to 23

21
- Flowers in distinct racemes
  Prunus virginiana demissa (choke-cherry), p. 110
- Flowers in small clusters, go to 22

22
- Leaves wedge shaped
  Prunus emarginata (bitter cherry), p. 110
- Leaves near heart shaped
  Prunus subcordata (Sierra plum), p. 110

23
- Leaves lack teeth
  Oemleria cerasiformis (oso berry), p. 108
- Leaves toothed on upper half
  Amelanchier pallida (service berry), p. 106

Encyclopedia of Shrubs

BARBERRY FAMILY (BERBERIDACEAE).

Mostly shrubs with pinnately compound leaves, leaflets spine edged and hollylike. Flowers in dense racemes, yellow; petals and sepals arrayed in series of 3s, stamens 6, single pistil with superior ovary. Fruit a grapelike berry.

California Barberry (*Berberis pinnata*).

Barberries are also known as holly grapes, since the leaflets of their compound leaves are prickly and closely resemble holly (but holly has simple leaves), and the purplish fruits are like miniature grapes. Sour when fresh, these fruits make excellent sauces and preserves. A trait unique to the barberries is the bright yellow inner bark, which served as a fine yellow dye among Native Americans. Barberries all have wandering roots, and so slowly—for they grow at a deliberate pace—establish large colonies, with many woody stems. California barberry is relatively rare in our area; consider yourself lucky when you run across it. One other rare barberry occasionally encountered is *B. dictyota* [*B. aquifolium dictyota*], with twisted grayish green leaves armed with exceedingly spiny teeth.

BEECH FAMILY (FAGACEAE).

Woody shrubs or trees, leaves variable, sometimes lobed, sometimes deciduous. Flowers unisexual; male flowers borne in long, dangling chains and consisting mainly of stamens; female flowers produced in small clusters and consisting of a cup of bracts (or spines) surrounding a single pistil and 3-lobed feathery stigma. Fruit a nut called an acorn.

Scrub Oak (*Quercus berberidifolia*). P. 298.

Our word for scrubby vegetation in California is "chaparral," from the Spanish word "chaparro" meaning scrub oak, since scrub oak typifies this kind of plant association. The ending "-al" indicates a grove or thicket of scrub oaks, certainly typical of the chaparral in southern California. Even so, chaparral is seldom of scrub oak alone, particularly in our area. Instead our local versions of chaparral have many other evergreen, drought-tolerant shrubs. Scrub oak is nearly always a shrub in our area, but in milder climates of southern California it may grow into multitrunked small trees. The tough, flat, toothed leaves are stiff and evergreen, paler beneath, and the bark is pale gray. The rather small acorns were considered of inferior quality for food, requiring relatively long to leach away the bitter tannins, and so were gathered only in years when live, California black, and valley oaks had poor acorn production.

Leather Oak (*Quercus durata*). P. 298.

Leather oak nearly matches scrub oak in stature and habit, but it favors especially nutrient-poor soils, such as those found on serpentine and volcanic outcrops. So constant is this association that leather oak is often considered a serpentine indicator. You can easily distinguish leather oak from a distance by the curled leaf edges and color: pale grayish green instead of dark green. Under a hand lens the reason for this color is obvious: hundreds of tiny, starburstlike whitish hairs cover the upper surface to reflect away the hot summer sun.

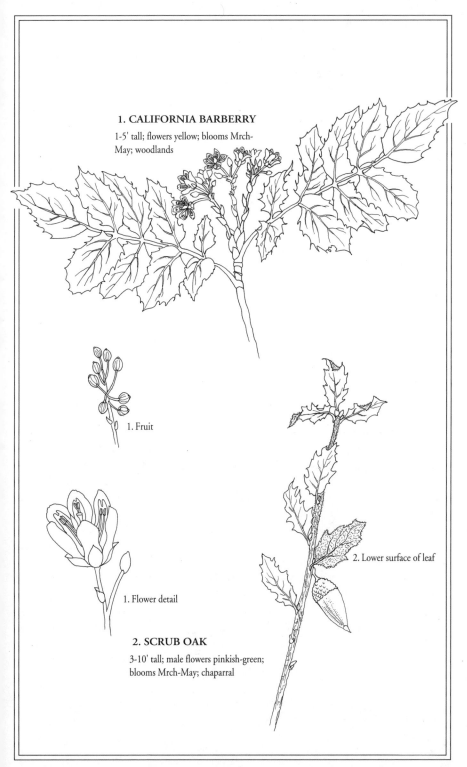

1. CALIFORNIA BARBERRY

1-5' tall; flowers yellow; blooms Mrch-May; woodlands

1. Fruit

1. Flower detail

2. Lower surface of leaf

2. SCRUB OAK

3-10' tall; male flowers pinkish-green; blooms Mrch-May; chaparral

Shrub Chinquapin (*Chrysolepis chrysophylla minor*). P. 297.

The word chinquapin is of Native American origin, while the scientific name tells us in no uncertain terms that this is "golden-scale golden-leaf." This golden quality is immediately obvious, for the tough, leathery, evergreen leaves are dark green above but covered with beautiful golden scales beneath. The last name appended—"minor"—tells us that this is a smaller version of the more usual coast chinquapin, which is a tree. The shrub form prefers rocky embankments, where it joins ranks with evergreen huckleberry and various manzanitas. The genus name formerly used by many books—"*Castanopsis*"—alludes to another interesting fact: this plant resembles the true chestnuts. ("Castanea" is Latin for chestnut; "opsis" means to look like.) Indeed, chinquapin is closely related to the true chestnuts and, like them, has upright catkins of white male flowers and spiny burrs enclosing the fruits. Here the resemblance ends, for the leaves are not chestnutlike with sharply serrated edges, and the fruits are small angled nuts, not the large round shiny seeds of true chestnuts. The fruits are edible, but two problems with their use for food are the formidable spines surrounding the nuts and the frequency with which insects lay their eggs inside the seeds before fruits ripen.

BUCKTHORN FAMILY (RHAMNACEAE).

Low to tall almost treelike shrubs, with simple leaves. Leaves variable, sometimes toothed, sometimes deciduous, sometimes opposite, sometimes with stipules. Flowers tiny, borne in dense, often complex clusters, sepals and petals colored alike. Five hooded sepals, five starlike to spoon-shaped petals, 5 stamens; single pistil with superior ovary. Fruit a 3-sided capsule or fleshy berry.

Ceanothuses or Wild Lilacs (*Ceanothus* spp.). P. 297.

The wild lilacs have great clusters of tiny white, pink, purple, or blue flowers that superficially look like lilacs. Instead the two are scarcely related, for true lilacs (genus *Syringa*) belong to the olive family Oleaceae and are native to Asia. A close encounter with ceanothus flowers quickly reveals a strong difference: ceanothus flowers have a sweet cornlike odor, while true lilacs have a powerful scent reminiscent of orange blossoms. Both plants use their fragrant flowers to attract bees; on warm spring days ceanothuses are loaded with bees busily mining the flowers for their rich nectar. Up close (with a hand lens), ceanothus blossoms are works of art: five hooded sepals match in color and alternate with five spreading, narrowly spoon-shaped petals. The stamens are attached to a pretty raised disc that secretes nectar.

Ceanothuses are important to the ecology of the chaparral, for their roots have tiny nodules in which nitrogen-producing bacteria live; when the roots die, these nutrients are released into soils to allow other shrubs to grow. Short lived by nature, ceanothuses are renewed from their roots after fire or from large numbers of seeds in the soil that await heat to crack the hard seed coats.

Four of our most typical ceanothuses are: buckbrush (*C. cuneatus*), with opposite, shiny, wedge-shaped leaves bearing pairs of wartlike stipules at their base and white to off-white flowers; blue blossom (*C thyrsiflorus*), a tall, almost treelike shrub with alternate, elliptical leaves having three main veins running their length and no stipules, and with trusses of blue-purple flowers; Jimbrush (*C. sorediatus* [*C. oliganthus sorediatus*]), a similarly large shrub with somewhat smaller but similar leaves, unangled twigs, and more rounded clusters of pale blue-purple flowers; and white-thorn (*C. leucodermis*), with large, dense sprays of palest lavender or white flowers on multitrunked shrubs with white bark and thorns. All occur in chaparral, but blue blossom generally prefers the fog belt while the other species are adapted to the full heat of summer.

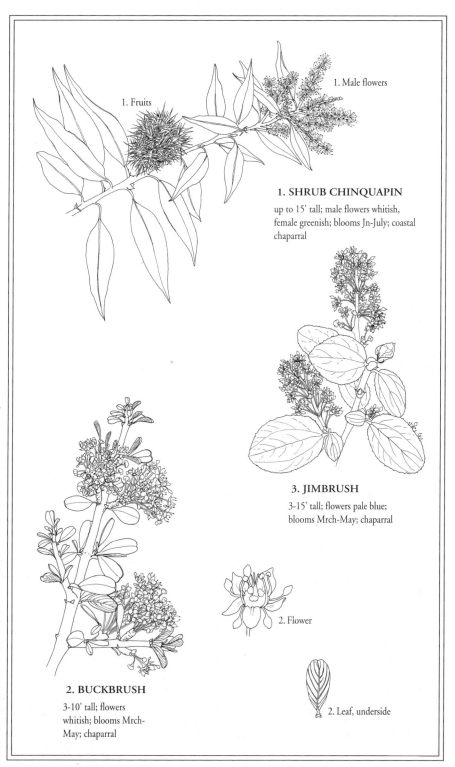

1. Fruits

1. Male flowers

1. SHRUB CHINQUAPIN

up to 15' tall; male flowers whitish,
female greenish; blooms Jn-July; coastal
chaparral

3. JIMBRUSH

3-15' tall; flowers pale blue;
blooms Mrch-May; chaparral

2. Flower

2. Leaf, underside

2. BUCKBRUSH

3-10' tall; flowers
whitish; blooms Mrch-
May; chaparral

Coffee Berry (*Rhamnus californica*) P. 298.

The reason for the common name coffee berry is not what it suggests, for this plant has no relationship to the beverage we so cherish. Instead, the laxative berries envelop two to three coffee-sized and -shaped seeds. The birds find these dark purple berries alluring and, in the process of eating them, inadvertently spread the seeds. Coffee berry is a variable plant standing less than a foot tall along the coast (where strong winds prune it low) to ten or more feet tall inland; leaves also vary from nearly flat and bright green to curled and dusky gray-green surfaces, according to the severity of the summer heat and drought. The pinnately arranged veins are distinctive; they arch out to just short of the leaf edge, and leaf margins have tiny sawlike teeth. Even in flower, coffee berry is not particularly distinguished, since the tiny, starlike flowers are yellow-green. Look for coffee berry in most shrubby communities, often in company with California sagebrush, ceanothuses, chamise, toyon, and huckleberry.

Redberry Buckthorn (*Rhamnus ilicifolia*). P. 298.

Although of the same genus as coffee berry, redberry buckthorn looks altogether different in its habit and leaves. Actually this is a highly variable shrub, with near prostrate forms along California's south-central coast and slender, tall shrubs (or small trees) inland around Mt. Diablo. The tough, evergreen leaves have a pretty sheen to them, bluish green when backlit. They are small, nearly-circular to oblong, and have a strong pinnate vein pattern easily seen on the underside. Leaf margins may be smooth or have prickly, hollylike teeth. Leaves of the latter form superficially resemble foliage of small-leafed live oaks, but the strong vein pattern immediately distinguishes them. Like coffee berry, redberry buckthorn has minute yellow-green starlike flowers, but the fall display of lustrous red berries is bright and eye-catching. Some shrubs are effectively male, others female, so that you may have to search awhile to find a bush with good berry display.

DAISY OR COMPOSITE FAMILY (ASTERACEAE).

Variable plants; flowers clustered together to resemble single large flower. Outer floral bracts resemble sepals, ray flowers on outside consist of single strap-shaped, showy petals and are often sterile; inner disc flowers are small, with 5 partly fused starlike petals, modified sepals called a pappus (a crown of hairs or scales at top of ovary), 5 partly fused stamens, and a single pistil with inferior ovary and 2 styles. Fruit a one-seeded achene.

California Sagebrush (*Artemisia californica*). P. 297.

California sagebrush is most closely related to the familiar Great Basin sagebrush that extends across vast areas of the Pacific Intermountain region in Nevada, Utah, and Arizona. In both types the intense sagelike odor drives away would-be diners, but in California sagebrush the leaves are finely wrought into narrow, linear divisions. Some leaves may be shed during particularly dry, hot summers, to be replaced after fall rains. The blossoms—demure heads of yellow-green disc flowers only—go nearly unnoticed at summer's end. Note that this sage-scented shrub is not related to the other sages (genus *Salvia*) of our area, the latter being members of the mint family Lamiaceae, but their similar scent adapts them to similar ecological niches.

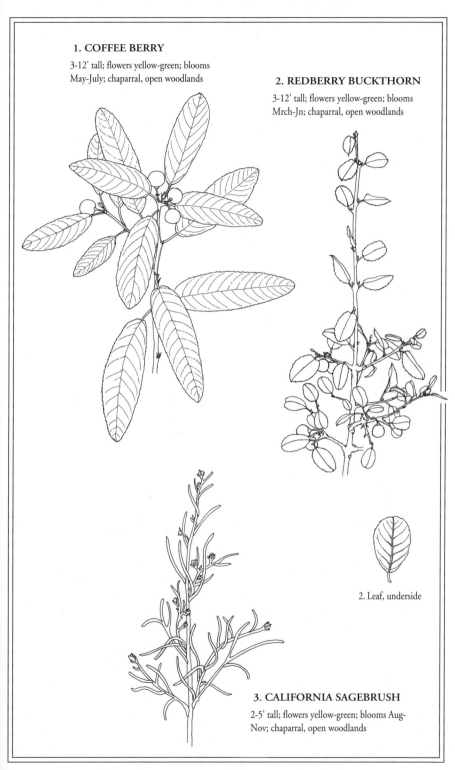

1. COFFEE BERRY

3-12' tall; flowers yellow-green; blooms
May-July; chaparral, open woodlands

2. REDBERRY BUCKTHORN

3-12' tall; flowers yellow-green; blooms
Mrch-Jn; chaparral, open woodlands

2. Leaf, underside

3. CALIFORNIA SAGEBRUSH

2-5' tall; flowers yellow-green; blooms Aug-
Nov; chaparral, open woodlands

Coyote Bush (*Baccharis pilularis*). P. 297.

Coyote bush is at once our most common shrub and most adaptable, being the first to move into newly cleared lands and vacant lots. This effective ability is owed to the white crown of hairs atop the fruits, which adapts them to wide-range dispersal by winds. Winds carry them long distances just as they do with the distantly related dandelions. Coyote bush is also noteworthy for its rapid growth, but later it is often crowded or shaded out by taller shrubs and trees. The small, elliptical leaves are shiny on warm days, and their fragrance then helps dissuade browsers. Evergreen all year, the only change in the shrubs' appearance is in fall when flower heads appear. Unlike most other composites, coyote bush has separate (dioecious) male and female plants. The male has small heads of cream-colored disc flowers, and the female has small heads of white disc flowers, the white coming from the long pappus hairs protruding beyond the other flower parts.

Mulefat (*Baccharis salicifolia*). P. 297.

Mulefat is distinctively different from coyote bush even at first glance; in place of the dense three-dimensional branches of the latter, mulefat throws up myriad vertical branches with narrow, lance-shaped, willowlike leaves. Growing in summer-dry streambeds and arroyos, mulefat may usurp the habitat of true willows, although the latter prefer a generally higher year-round water table. Only in summer does mulefat reveal its relationship to coyote bush. Then its small male and female flower heads of disc flowers appear, although the differentiation between the sexes is not so strong; both bear disc flowers with tiny white petals.

California Brickel Bush (*Brickellia californica*). P. 297.

California brickel bush is a seldom noticed component of our chaparral, particularly along the edge of summer-dry arroyos. Look for soft gray-green leaves with a triangular outline that are purplish when new. On hot days, the leaves have a pleasant, sweet odor that is not obvious on cool days. The clusters of small, greenish flower heads in fall are a disappointment, but if you wait until late afternoon or evening, you'll smell a potent sweet perfume wafted around by breezes. This perfume is intended for night-flying insects that aren't attracted by brightly colored flowers but find powerful allure in the promise of that perfume.

Bush Sunflowers (*Ericameria linearifolia* and *arborescens*). P. 297.

Out of blossom, our two bush sunflowers look rather similar: medium-sized shrubs with dense branches and a plethora of pine-scented, narrowly linear, resinous bright green leaves. The habit differs on mature plants, for *E. arborescens* (golden fleece) has a distinct trunk and top crown of branches, whereas *E. linearifolius* (narrow-leaf bush sunflower) forms a dense, multibranched shrub. Both like rocky slopes in full sun; hot summers don't faze these stalwart survivors. In flower the two differ: bush sunflower has myriad bright yellow daisies (often an inch or more across and nearly hiding the rest of the bush) in mid-spring, while golden fleece produces scattered clusters of small rayless yellow flower heads in late summer or fall. Both are transformed a second time when the white-hairy pappuses of the fruits are displayed after flowering.

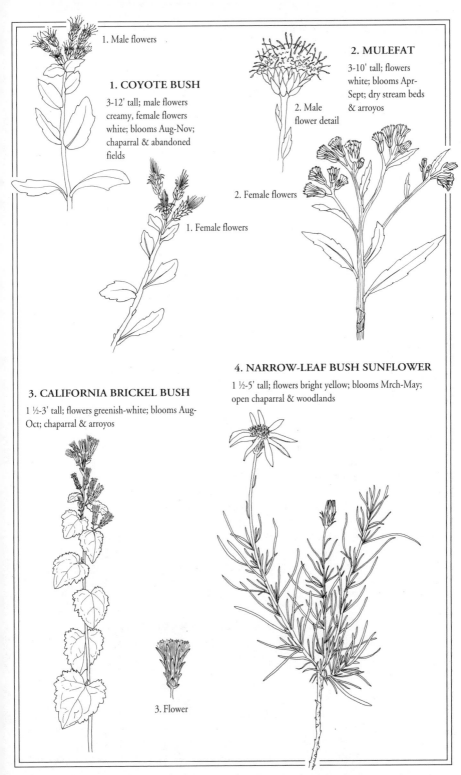

1. Male flowers

1. COYOTE BUSH

3-12' tall; male flowers
creamy, female flowers
white; blooms Aug-Nov;
chaparral & abandoned
fields

2. MULEFAT

3-10' tall; flowers
white; blooms Apr-
Sept; dry stream beds
& arroyos

2. Male
flower detail

2. Female flowers

1. Female flowers

4. NARROW-LEAF BUSH SUNFLOWER

1 ½-5' tall; flowers bright yellow; blooms Mrch-May;
open chaparral & woodlands

3. CALIFORNIA BRICKEL BUSH

1 ½-3' tall; flowers greenish-white; blooms Aug-
Oct; chaparral & arroyos

3. Flower

DAPHNE FAMILY (THYMELEACEAE).

Shrubs or small trees with simple leaves. Flowers variable in shape; 4 or 5 colored tepals, 4 or 5 stamens, and a single pistil with inferior ovary. Fruit a drupe or berry.

Western Leatherwood (*Dirca occidentalis*). P. 297.

Harbinger of spring, western leatherwood is our only native member of the daphne family. It is also, along with several species from the Old World, the only well-known member of the family. Our leatherwood is doubly special since it is so restricted in distribution, being found only along the interface between mixed-evergreen forest and soft chaparral in counties around the Bay. Its only close relative appears to be an eastern species of leatherwood, a couple of thousand miles away! Look for the nodding, bright yellow bells at winter's end, just as other shrubs are gaining new leaves or the pink flowering currant has burst into beautiful bloom. The soft, pale green, broadly elliptical leaves appear just after blossoms fade. Once leafed out, leatherwood resembles nothing so much as oso berry, a small shrub in the rose family occurring in similar habitats. At that time, leatherwood may be distinguished by the highly flexible twigs that can be twisted into a circle without breaking.

DOGWOOD FAMILY (CORNACEAE).

Shrubs or trees with red new twigs, pairs of broadly elliptical deciduous leaves with distinctive vein pattern, and dense clusters of starlike white flowers. Our species lacks the showy, petallike floral bracts of other dogwoods and has flowers in flat-topped cymes; 4 minute sepals and 4 separate petals, 4 stamens, single pistil with inferior ovary. Fruit a few-seeded berry or drupe.

Creek Dogwoods (*Cornus glabrata* and *occidentalis* [*C. sericea occidentalis*]). P. 297.

Creek dogwoods are poorly known relatives of the beautiful flowering dogwood; the latter is common in the mountains but not found in our region. Creek dogwoods form dense, multibranched hedges along permanent watercourses, sometimes under willows or other taller trees or shrubs. Deciduous, their vivid red-brown to red-purple twigs stand out in winter; in spring and summer they wear a fine greenery of broad leaves; and in autumn these leaves turn lovely shades of purple, pink, or yellow before falling. The most distinctive difference between them and flowering dogwood is that the latter has showy white petallike bracts surrounding its tight button of flowers; creek dogwood's flowers are unprepossessing by comparison, in loose clusters, each flower a perfect miniature white star followed by white to bluish fruits. The two species are closely similar: *C. glabrata* (brown dogwood) has nearly hairless leaves and sometimes brownish twigs; *C. occidentalis* (western creek dogwood) has densely felted hairs underneath its leaves and bright red twigs.

FIGWORT FAMILY (SCROPHULARIACEAE).

Sometimes small shrubs with simple, opposite leaves and showy flowers; flowers irregular, 2-lipped; 5 partly fused sepals, 5 partly fused petals, 4 or 5 stamens, single pistil with superior ovary and 2-lobed stigma. Fruit a capsule.

Bush or Sticky Monkeyflower (*Diplacus* [*Mimulus*] *aurantiacus*). P. 297.

The bush monkeyflowers are familiar small showy shrubs of our chaparral communities: typically a permanent member of the soft chaparral and a temporary member of the hard chaparral, especially after fire. Common companions include coffee berry, California sagebrush,

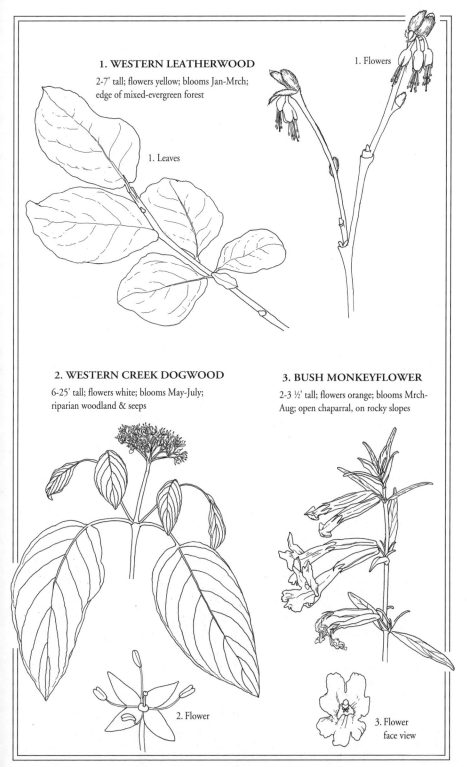

1. WESTERN LEATHERWOOD

2-7' tall; flowers yellow; blooms Jan-Mrch; edge of mixed-evergreen forest

1. Flowers

1. Leaves

2. WESTERN CREEK DOGWOOD

6-25' tall; flowers white; blooms May-July; riparian woodland & seeps

2. Flower

3. BUSH MONKEYFLOWER

2-3 ½' tall; flowers orange; blooms Mrch-Aug; open chaparral, on rocky slopes

3. Flower face view

and coyote bush. On hot days the leaves produce a sticky varnish to cut down on water loss, but if the summer days remain hot and dry for long the bushes begin the process of defoliation, recovering quickly after winter rains. The best feature is the showy, bright orange flowers ("aurantiacus" means orangey), borne in profusion at spring's end and continuously through summer if the shrub is grown in a garden with supplemental water. Start new shrubs from tip cuttings in spring. Each flower is two-lipped, the throat hiding a pair of long and short stamens with orange pollen, but the two-lobed stigma extends out to the entrance to the throat, where a bee will brush against it in its quest for nectar inside the throat. Within seconds after contact, the two white flaps of the stigma have closed, assuring that a new load of pollen from the same flower cannot accidentally get deposited on its own stigma.

Gaping Penstemon (*Keckiella breviflora*). P. 298.

The distinctly woody, shrubby penstemons have been reclassified as keckiellas, although they're closely related to the true penstemons. Like the latter, their flowers are distinguished by four fertile stamens and one sterile stamen. This sterile "staminode" is bearded with bushy hairs in the keckiellas (another common name for this fifth stamen is bearded tongue). Gaping penstemon is undistinguished out of flower—a lank low-growing shrub with greenish twigs and small triangular toothed leaves. But the flowers bear close inspection for their unusual shape and beauty. The white petals gape wide as though the two lips of the blossom were yawning, and the lower lip is prettily striped with long, purple nectar guide lines.

Red Rock Penstemon (*Keckiella corymbosa*). P. 298.

Rare with us, red rock penstemon—"red" for the flowers—clings to hot, exposed sandstones and cherts near the top of Mt. Diablo. Though sprawling close to the ground up there, the plant may grow upright in less windy situations. Look for the brilliant red flowers with long tubes and widely spreading petals in summer, when the flowers provide a source of nectar for hummingbirds. Like so many penstemons, this one is a great drought-tolerant shrub for the garden and a star performer during the long dry season.

GARRYA FAMILY (GARRYACEAE).

Tall shrubs or small trees with opposite, leathery, simple, evergreen leaves. Plants dioecious: male flowers in long, dangling chains consisting of sets of floral bracts and protruding bunches of stamens; female flowers in shorter chains with similar bracts and single pistils with protruding stigmas. Fruit a berry with red-purple pulp.

Silk Tassel Bush (*Garrya* spp.). P. 297.

Out of flower, silk tassel bushes may be confused with manzanitas, for their tough, evergreen leaves are shaped in similar fashion. But they're arranged in pairs instead of alternately, and they're often wavy as well. In addition, silk tassel bushes have grayish-brown bark, while manzanitas are easily told by their smooth, red-purple bark. In flower, there's a world of difference, for the wind-pollinated gray-green flowers are arranged in long chains and appear in the middle of winter, when winds are most reliable. In the garden, they make handsome, drought-resistant shrubs, but the male shrub is always chosen since its tassels are longer and showier than those of the female. Our two species include *G. elliptica* (coast silk tassel), with dark green wavy leaves. It is frequent in the fog belt. *G. fremontii* (Fremont's silk tassel), with bright green smooth leaves, is more typical of the chaparral inland, beyond the fog belt.

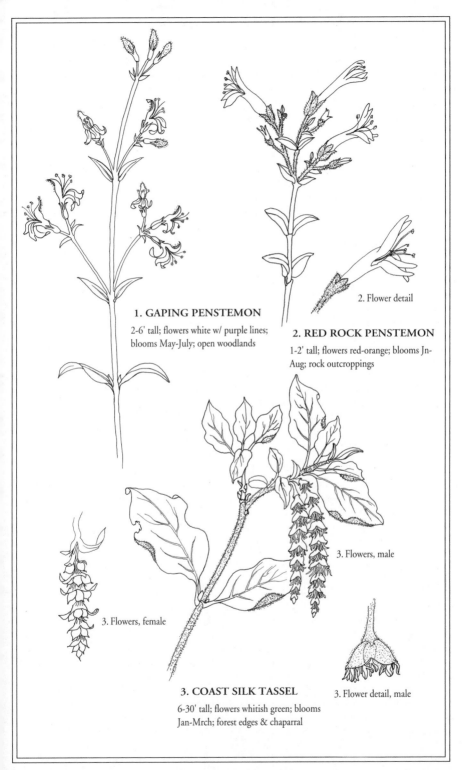

1. GAPING PENSTEMON

2-6' tall; flowers white w/ purple lines; blooms May-July; open woodlands

2. Flower detail

2. RED ROCK PENSTEMON

1-2' tall; flowers red-orange; blooms Jn-Aug; rock outcroppings

3. Flowers, male

3. Flowers, female

3. COAST SILK TASSEL

6-30' tall; flowers whitish green; blooms Jan-Mrch; forest edges & chaparral

3. Flower detail, male

GOOSEBERRY FAMILY (GROSSULARIACEAE).

Mostly deciduous shrubs with palmately lobed, alternate leaves. Flowers hanging in axillary clusters or in dense racemes; 4 to 5 sepals are fused to form a tube (and are often showier than the petals). Four to 5 separate petals, 4 to 5 stamens, single pistil with inferior ovary. Fruit a berry.

Flowering Currants (*Ribes* spp.). P. 298.

The flowering currants will never win favor for their bland fruits, but they're among our most ornamental native shrubs in flower—usually from late winter to early spring. The pretty leaves are shaped like miniature maples, but they are sticky on warm days and scented with the odor of sage. Trusses of rose, pink, or pale red-purple flowers appear before the new leaves have fully developed. The flowers themselves are curious, for the flaring pink sepals are larger and more obvious than the narrow, cuplike petals. Our two common species are pink flowering currant (*Ribes sanguineum glutinosum*), with pink flowers in fog-belt forests, and chaparral currant (*R. malvaceum*), with pale red-purple flowers in open woodlands and chaparral. Much rarer here is the golden currant (*R. aureum*), with bright yellow flowers. All make handsome specimens in the lightly shaded garden, as under oaks.

Gooseberries (*Ribes* spp.). P. 298.

Gooseberries differ in several particulars from currants: their leaves tend to be smaller, less resinous or sticky; the flowers are borne in modest clusters in leaf axils and hang; the stems are armed with three spines per node; and often the fruits are covered with glandular hairs and/or spines. Again, the commercial gooseberries are not from local sources, and most of ours are unremarkable for their edibility, but the jewellike flowers in early spring are especially lovely at close range. Three gooseberries are especially characteristic of our area. *R. menziesii*, the canyon gooseberry, has horizontally tiered branches and charming flowers combining rich maroon-red sepals with cupped white petals, a bit like miniature fuchsias. *R. californicum*, the hillside gooseberry, is similar but lacks the brushy internodal spines of the canyon gooseberry. *R. quercetorum*, the oak-leaf gooseberry, is a straggly shrub with drooping branches and rather obscure yellow and greenish blossoms. The latter is abundant in the Livermore hills, while canyon and hillside gooseberries are widespread in moist woodlands. One outstanding but rare gooseberry barely entering our area is *R. speciosum*, the fuchsia-flowered gooseberry. This has long, tubular scarlet flowers in spring—attractive to hummingbirds—and makes a handsome addition to any woodland garden.

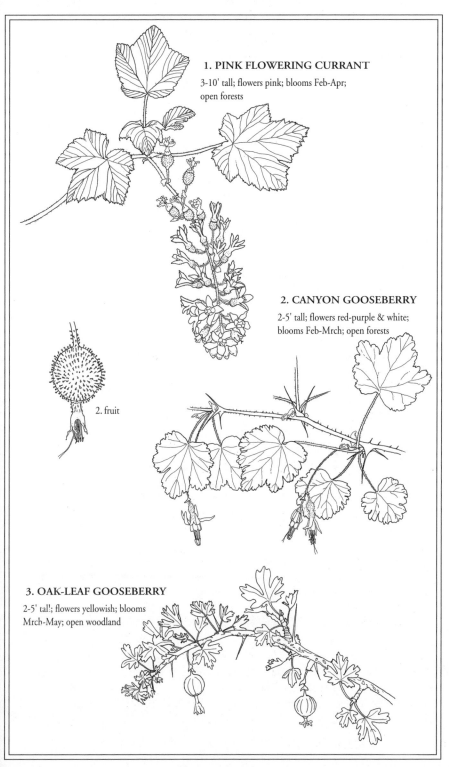

1. PINK FLOWERING CURRANT

3-10' tall; flowers pink; blooms Feb-Apr;
open forests

2. CANYON GOOSEBERRY

2-5' tall; flowers red-purple & white;
blooms Feb-Mrch; open forests

2. fruit

3. OAK-LEAF GOOSEBERRY

2-5' tal!; flowers yellowish; blooms
Mrch-May; open woodland

HEATHER FAMILY (ERICACEAE).

Woody plants with simple, often leathery leaves. Flowers mostly bell to urn shaped; 5 sepals, 5 fused petals (the lobes often quite small), 5 or 10 stamens with tiny holes at tips of anthers (and often accompanied by turned-down "horns"). Single pistil usually with superior ovary. Fruit a berry or capsule.

Manzanitas (*Arctostaphylos* spp.). P. 297.

This marvelously varied group of shrubs is typically Californian, extending from the seashores to the highest mountain forests, and is especially prolific under the harsh conditions found in chaparral. Although ranging from low, creeping shrubs to near-trees up to twenty feet tall, all manzanitas share these characteristics: smooth (usually), red-purple bark; simple, untoothed ovate, leathery, evergreen leaves; and nodding racemes or panicles of white or pink urn-shaped flowers in late winter to early spring. The common name is from Spanish for "little apple," another genus-wide trait, for the pink, rose, or mahogany-brown fruits resemble nothing so much as little apples. They are mealy inside and contain one to several bone-hard stones. The Native Americans were especially fond of ripe manzanita berries; berries are also excellent for manzanita tea or jelly. But the wild animals that flock to them (the generic name is Greek for bear grape) are the real benefactors to the shrubs, for the hard seed coats are softened by passing through the gut, ready for germination when they are excreted. Today, manzanitas create a wonderful centerpiece for the drought-tolerant garden or serve as low-maintenance ground covers.

The splitters and lumpers have had a heyday in trying to resolve the question of the enormous variation found in manzanitas; this is further complicated by the species' gregarious gene exchange. Our area is noted for several species, some of which are highly restricted endemics. Two such are pallid manzanita (*A. pallida*), a near-tree-sized species with gorgeous smooth-barked trunks and pale pinkish flowers, from the Oakland-Berkeley hills, and Mt. Diablo manzanita (*A. auriculata*), a localized shrub on Mt. Diablo with softly felted, gray, clasping leaves and pink flowers. Other species include common manzanita (*A. manzanita*), a large shrub to small tree with green, stalked leaves and large trusses of white or pale pink flowers appearing as early as winter; brittle-leaf manzanita (*A. crustacea* [*A. tomentosa crustacea*]), a small shrub with basal burl and especially stiff gray-green leaves and white flowers; and big-berry manzanita (*A. glauca*), a small tree with lovely whitish gray leaves, white flowers, and especially large sticky reddish berries.

Evergreen Huckleberry (*Vaccinium ovatum*). P. 299.

Common in the redwood belt, evergreen huckleberry continues northward into the great conifer forests of the Pacific Northwest. Huckleberry often forms near-impenetrable hedges along the edge of such forests; closer to the coast bushes grow taller than those in the Oakland hills. The dark brown bark and small, horizontal, shiny green, toothed leaves little resemble those of manzanitas, but the bell-shaped white to pink blossoms are close approximations, although they hang almost hidden beneath leafy branches. The most memorable phase of the life cycle is the profusion of pale purple to deep blue-purple berries that are ripe by summer's end and delicious by any standard. These are fantastic in just about any recipe calling for blueberries, including pies, muffins, and pancakes. Berries with a whitish "bloom" on them are actually covered with naturally occurring native yeasts, and such berries are outstandingly sweet.

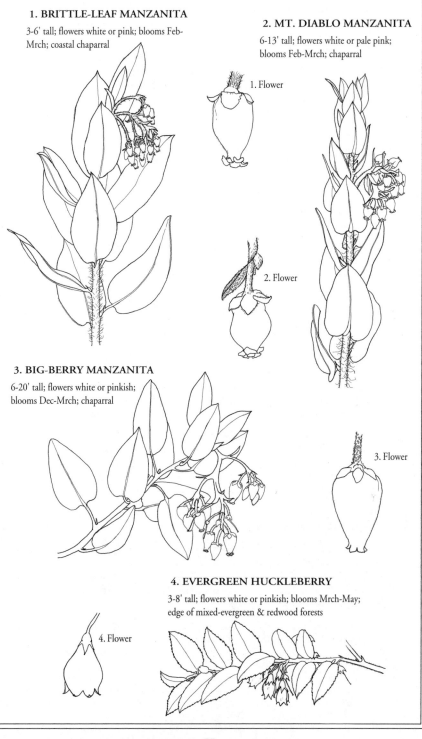

1. BRITTLE-LEAF MANZANITA

3-6' tall; flowers white or pink; blooms Feb-Mrch; coastal chaparral

2. MT. DIABLO MANZANITA

6-13' tall; flowers white or pale pink; blooms Feb-Mrch; chaparral

1. Flower

2. Flower

3. BIG-BERRY MANZANITA

6-20' tall; flowers white or pinkish; blooms Dec-Mrch; chaparral

3. Flower

4. EVERGREEN HUCKLEBERRY

3-8' tall; flowers white or pinkish; blooms Mrch-May; edge of mixed-evergreen & redwood forests

4. Flower

HONEYSUCKLE FAMILY (CAPRIFOLIACEAE).

Woody shrubs or vines with opposite, simple to compound leaves. Flowers borne in varied arrangements, often individually small; 5 minute sepals, 5 partly fused petals (sometimes irregular), 5 stamens fused at base to petals, single pistil with inferior ovary. Fruit a berry.

Twinberry Honeysuckle (*Lonicera involucrata ledebourii*). P. 298.

The usual viny honeysuckles, as exemplified by the cultivated Japanese honeysuckle, are quite different in appearance from our shrub twinberry honeysuckle. This medium-sized shrub seeks permanently wet habitats, such as seeps and streamsides in open forests, where it leafs out in early spring, then flowers and fruits. A pair of red floral bracts accentuates the pair of slender yellow-orange blossoms, making them attractive to hummingbirds. But the real show comes later when each blossom is replaced by a large shiny black berry, beautifully set off by those same red bracts. Although the fruits leave a foul aftertaste, birds are fond of them.

Snowberry (*Symphoricarpos albus laevigatus*). P. 298.

Snowberry is a small shrub of our forests and woodlands with considerable seasonal interest. In winter, it's identified by the large colonies (roots creep to send up ever more shoots) and the exceptionally thin, fine twigs. In spring, pairs of pale green leaves reappear and hide the tiny pinkish bell-shaped flowers at spring's end. In summer, the shrubs doze. And in fall, the flowers have given way to short clusters of spongy, snowy white berries. These berries are insipid, and said by some to have a soapy quality, but as with twinberry honeysuckle the birds find them appetizing. Without flowers or fruits, snowberry is identified by a peculiar attribute: some leaves are even and untoothed, others look like someone took a large bite out of one side, still others have shallow lobes or scallops. Although *S. albus* is our only shrubby species, there is a creeping, ground-hugging snowberry occasional in our woodlands: *S. mollis*.

MALLOW FAMILY (MALVACEAE).

Some small shrubs with greenish twigs and palmately lobed to rounded leaves. These have palmate veins and tiny stipules, and starlike hairs on stems as well as leaves. Showy, cup-shaped flowers; 5 sepals interspersed with 5 sepallike floral bracts, 5 separate petals, numerous stamens fused into a tube, single pistil with superior, multichambered ovary. Fruit consists of ovary sections separating like slices from a cheese wheel.

Bush Mallows (*Malacothamnus* spp.). P. 298.

Although they're components of our chaparral, bush mallows are little known, for they are rare most of the time. The exception is after fire, when the long-dormant seeds are stimulated to germinate by the thousands. For a span of a few years, bush mallows may dominate the landscape as they quickly grow to maturity and blossom prolifically. Most species of bush mallow occur to the south of us, but *M. fremontii* has been especially prominent on Mt. Diablo after fires. This attractive shrub has closely felted gray leaves, with thick starry hairs beautiful to see under a hand lens. The pale lilac flowers complement the leaves and are in turn complemented by the yellow-orange stamens.

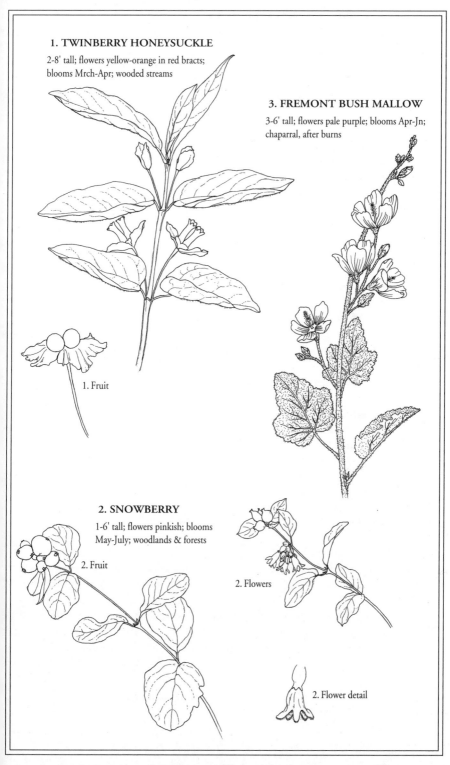

1. TWINBERRY HONEYSUCKLE

2-8' tall; flowers yellow-orange in red bracts; blooms Mrch-Apr; wooded streams

3. FREMONT BUSH MALLOW

3-6' tall; flowers pale purple; blooms Apr-Jn; chaparral, after burns

1. Fruit

2. SNOWBERRY

1-6' tall; flowers pinkish; blooms May-July; woodlands & forests

2. Fruit

2. Flowers

2. Flower detail

MINT FAMILY (LAMIACEAE).

Some are small shrubs with pairs of fragrantly scented leaves borne on square stems. Flowers often in whorls, irregular and 2-lipped; 5 mostly fused sepals, 5 partly fused petals, 2 above and 3 below. Two or 4 stamens, single pistil with superior ovary and 2-lobed stigma. Fruit, 4 one-seeded nutlets.

Pitcher Sage (*Lepechinia calycina*). P. 298.

Look for this sage look-alike on the edge of chaparral, after burns, or among the smaller shrubs of the coastal scrub. At first glance, it looks like a true sage (genus *Salvia*), but the softly felted and quilted leaves have a special odor that many may find offensive—strong and distinctive in any case. The horizontally held lines of flowers are quite pretty, with five prominent, clearly net-veined sepals surrounding a short petal tube and two flared lips, the latter in pale purple or white. The most distinctive feature of pitcher sage is in the namesake pitcherlike shape of the sepals that gradually inflate around the four marble-shaped fruits inside. Pitcher sage is most often encountered in clearings or burned areas as a short-lived component of chaparral.

Black Sage (*Salvia mellifera*). P. 298.

Black sage is our only true shrubby sage, but its leaves are far more strongly scented than those of the Mediterranean culinary sage (*S. officinalis*). In fact, each species of salvia has its own special odor, some overbearing, others sweet, fruity, subtle, or bold. Perhaps black sage is named for the dark green color of the leaves, which when massed, from a distance give a dark appearance to the slopes on which they grow. Although the books refer to them as evergreen, in fact black sage often loses large numbers of leaves during prolonged summer droughts. Sometimes occurring in exclusive stands, black sage is particularly well adapted to some of our hottest, rockiest slopes, but there may be another reason for the apparent lack of competition. Volatile oils—the reason the leaves smell strongly on hot days—may inhibit the growth of surrounding vegetation, particularly newly sprouted, competing seedlings. In flower, black sage has several tiers of modest, pale purple blossoms irresistible to bees, from which a fine honey is made. This gives us the species name "mellifera," which means honey producing.

NIGHTSHADE OR POTATO FAMILY (SOLANACEAE).

Some are small, green-woody shrubs with ill-smelling, simple leaves. Flowers showy, often in umbels; 5 partly fused sepals, 5 much-fused petals, 5 stamens often adhering together and fused at their bases to petals, single pistil with superior, 2-chambered ovary. Fruit a capsule or berry.

Tree Tobacco (*Nicotiana glauca*).

Native tobaccos extend all the way from western North America to southern South America. Its southern homeland produced the smoking tobacco, *N. tabacum*. Of the nicotianas, most are annual or perennial herbs, but a few grow to shrub size or even the stature of small trees. Such is the case with this species, originally from Brazil and constantly extending its range into California's dry arroyos and gravelly roadsides. Evidently as the seeds fall to the ground they get carried in the gravel around them, whether by automobile tires, road working equipment, or other means. Tree tobacco is immediately recognizable by its tall, wandlike stems, with few flexible high-reaching main branches and bluish green leaves. The flowers are quite beautiful, being long, slender, pale yellow tubes, fluted at their ends where petal lobes diverge. The flowers are eagerly visited by hummingbirds, just as they are in their original homeland.

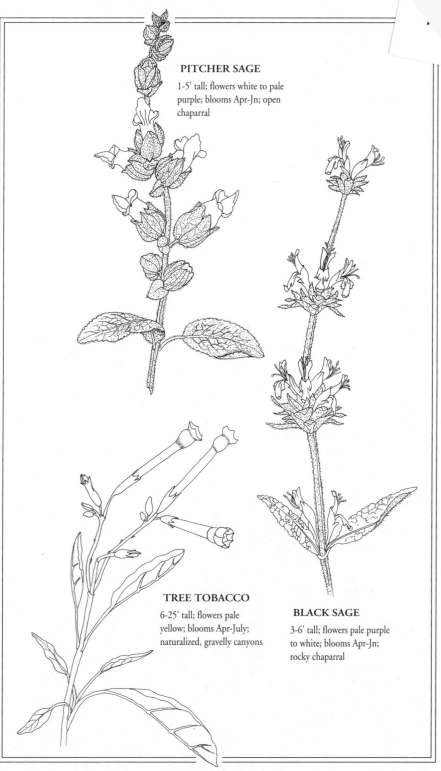

PITCHER SAGE

1-5' tall; flowers white to pale
purple; blooms Apr-Jn; open
chaparral

TREE TOBACCO

6-25' tall; flowers pale
yellow; blooms Apr-July;
naturalized, gravelly canyons

BLACK SAGE

3-6' tall; flowers pale purple
to white; blooms Apr-Jn;
rocky chaparral

lue Witch or Nightshade (*Solanum umbelliferum*). P. 298.

Blue witch is another of our small chaparral shrubs, taking advantage of the chaparral edge, road cuts, burns, and temporary clearings. Its many slender, green-felted twigs may be bare in summer when leaves are shed due to drought, but the leaves reappear after rains. In flower the shrub is transformed, for the inch-wide bright blue-purple saucers nearly smother it at peak bloom. Each flower has five points along the rim of the saucer, and each petal is trimmed at its base by a green nectar gland rimmed in white, the whole complemented by projecting bright yellow shooting-starlike stamens. When the fruits are young, they look like miniature green tomatoes (including the seeds inside), but they gradually turn to a dark black-purple as they ripen. Whether they're poisonous at this stage is not certain, for in any event the birds that feed on them are immune, but this genus has a reputation for possessing poisonous alkaloids throughout the plant.

OLIVE FAMILY (OLEACEAE).

Trees or shrubs, often with deciduous, pinnately compound leaves. Flowers small, usually borne in dense clusters; 5 partly fused sepals, 5 partly fused petals (sometimes missing), 2 stamens fused at base to petals, single pistil with superior ovary. Fruit a capsule or winged samara.

Flowering Ash (*Fraxinus dipetala*). P. 297.

Flowering ash is a modest, shrublike small tree, favoring canyon bottoms with a permanently high water table. Other ashes are similar in their habitat, but the species that occur on more permanent streams or rivers are larger trees and are not found here. Fresno ash and Oregon ash are two such species; they also differ in having longer, pinnately compound leaves, and unisexual, wind-pollinated greenish flowers borne on separate trees. By contrast, flowering ash has attractive nodding clusters of white flowers—each flower with two petals, giving the species name "dipetala"—and small leaves, divided normally into only three leaflets (sometimes five). In autumn when leaves have fallen, you can still recognize flowering ash by its hanging chains of samaras, but unlike maples with their doubly winged samaras, each ash samara has but one slender wing. These wings are useful in carrying the fruits on strong winds.

PEA FAMILY (FABACEAE).

Some are small shrubs, mostly with compound leaves and stipules. Flowers irregular, pea shaped. There are 5 partly fused sepals; 5 petals, the upper one a banner; 2 side petals called wings; 2 bottom middle petals that fuse into a boat-shaped keel; usually 10 stamens, 9 fused by their filaments, 10th separate; a single pistil with one-chambered superior ovary. Fruit a pea pod (legume).

Brooms (*Cytisus, Genista* and *Spartium* spp.).

Brooms are long-valued ornamental shrubs throughout California, but people are now becoming alarmed by their potential for harm to our native ecosystems. Some species and cultivars are well behaved. These maintain their place in the garden, not escaping to proliferate in the wilds. But the most commonly cultivated species have created drastic ecological problems. A trio of species is mostly responsible—French, Spanish, and Scotch brooms, closely similar shrubs that grow rapidly, flower profusely, and seed in great abundance. Seeds may remain viable in the soil for several years. The reason that brooms are such bad news is that their invasive abilities have allowed them to move along roadsides, especially in wooded areas, where they eliminate diversity in the flora and the fauna. In flower, it's easy to forget these faults, for the many green-twiggy branches are showered with golden yellow pea blossoms of

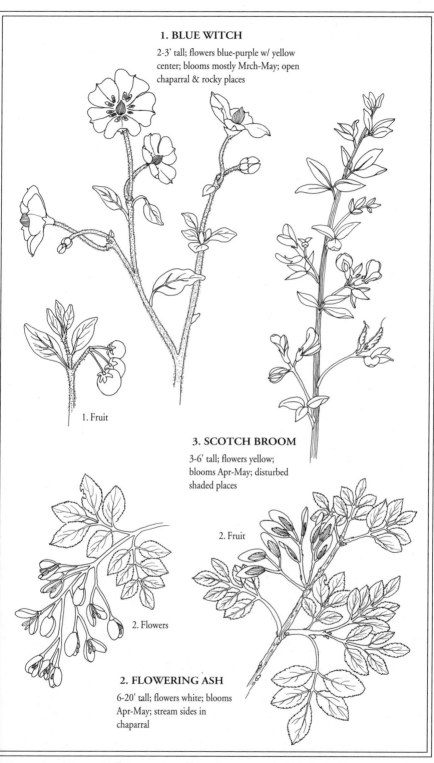

1. BLUE WITCH

2-3' tall; flowers blue-purple w/ yellow
center; blooms mostly Mrch-May; open
chaparral & rocky places

1. Fruit

3. SCOTCH BROOM

3-6' tall; flowers yellow;
blooms Apr-May; disturbed
shaded places

2. Fruit

2. Flowers

2. FLOWERING ASH

6-20' tall; flowers white; blooms
Apr-May; stream sides in
chaparral

great beauty. One other "alien" shrub deserves mention here: gorse (*Ulex europaea*) is a broomlike European bush heavily armed with green spines in place of leaves. Fortunately it seldom proliferates in our region, but it has become an invasive pest in heavily grazed pastures along the coast.

Blue Bush Lupine (*Lupinus albifrons*). P. 298.

Lupines span more than sixty species in California; most are nonwoody plants discussed elsewhere. A few are small shrubs adapted to sandy or rocky soils, where the nitrogen-fixing nodules on their roots help them enrich the soil and survive where other plants cannot. Like the nonwoody lupines, bush lupines are immediately identified by the palmately compound leaves, the leaflets that are splayed like the fingers on a hand, and the spikelike clusters of showy pea-shaped flowers. Blue bush lupine combines complementary silvery-haired leaves with blue-purple flower spires, and heightens their beauty by releasing a sweet perfume in the afternoon. Soon the floral show is over, and the bush may partially defoliate through the long, dry days of summer. Blue bush lupine is the only reasonably well represented shrub lupine in our area. Look for yellow tree lupine (*L. arboreus*) near the coast. It is occasionally planted here, beyond its natural range.

Chaparral Pea (*Pickeringia montana*). P. 298.

Of all the chaparral shrubs, only chaparral pea is a common component from the farflung pea family. Like its cousins, chaparral pea has a competitive edge on rocky soils, where its nitrogen-fixing nodules help roots find nourishment. Chaparral pea seldom sets viable seeds, and new colonies occur only rarely, but once a suitable habitat has been found, the plants quickly take advantage by sending out myriad roots that sprout new shoots as they go. Soon a whole hillside may be covered by one original colonist. Out of flower, the three small, dark green leaflets are not particularly eye-catching, but a close encounter with this shrub is not forgotten, for the branches end in stout thorns. In blossom, this shrub is transformed into a vision of true beauty: thousands of magenta pea flowers cover every branch. Such a vivid color is unusual in our flora, and since seed is sparsely set, one has to wonder what pollinator this floral display is meant to attract.

POPPY FAMILY (PAPAVERACEAE).

Some are small shrubs with simple leaves. Flowers showy, regular; 2 or 3 sepals that fall as flower opens, twice the number of petals, numerous stamens, single pistil with superior ovary. Fruit a capsule.

Bush Poppy (*Dendromecon rigida*). P. 297.

Sprinkled through the chaparral, bush poppy lights up the scene when covered with bright yellow, saucer-shaped flowers. During mild winters and summers, flowers may continue but are sparsely produced. The obliquely held, willowlike leaves are stiff to resist wilting and to avoid the full impact of hot summer sun. The flowers are followed by elongated seed pods that curl open from the bottom to the top in several pieces to liberate seeds. Bush poppy is only truly common after fires, when seeds germinate by the thousands, or along the borders of chaparral where it can compete with lower growing shrubs. Common companions include chaparral pea and yerba santa.

BLUE BUSH LUPINE

2-5' tall; flowers blue-purple; blooms
Mrch-May; open woodlands &
rocky slopes

CHAPARRAL PEA

2-8' tall; flowers magenta;
blooms May-Jn; edge of
chaparral & after burns

BUSH POPPY

3-10' tall; flowers bright yellow;
blooms Apr-Jn; edge of
chaparral & after burns

ROSE FAMILY (ROSACEAE).

Variable plants; many are shrubs and small trees. Leaf shape variable, leaves often with a pair of tiny stipules at the base. Flowers roselike, arranged in many fashions; 5 usually-cup-shaped sepals, 5 separate petals, usually numerous spirally arranged stamens, pistils single to many, with superior to inferior ovaries (or in between). Fruits achenes or fleshy pomes, aggregate fruits, or drupes.

Chamise (*Adenostoma fasciculatum*). P. 297.

Of all the chaparral shrubs, chamise is the most widespread and most drought tolerant, often occurring in pure stands on the rockiest, driest slopes, and at other times mingled with a variety of other shrubs: ceanothuses, silk tassel bush, scrub oak, mountain mahogany, manzanitas, or toyon. Ever adaptable, chamise may occur on volcanic soils, soils derived from slates, schists, and sandstones, or even serpentine soils. This shrub is easily identified by its dense clusters of needlelike leaves that turn oily and shiny on hot days. This makes chamise vulnerable to fire (and earns for it another common name: greasewood). In late May to early June, the shrubs are transformed by myriad spirelike clusters of creamy flowers, each flower like a minute single rose. Later, the rust-colored fruits give the effect of a brownish mantle from a distance.

Service Berry (*Amelanchier pallida* [*A. utahensis*]). P. 297.

Service berry is a variable shrub in a genus noted for its wide distribution across the country; with us it's relatively rare, occurring scattered sporadically on protected canyon slopes or along the edge of woods. Whether you call it other names, such as sarvice, June berry, or shad, this genus has enjoyed wide use for its miniature red-purple pomes, fruits organized along the lines of scaled-down apples. Fruits were widely dried by the Indians for mixing with bear fat as a travel food called pemmican, and pioneers often made jelly from them. Perhaps the best identifying trait for these large shrubs is their pale grayish-green deciduous leaves that bear saw teeth along the upper one-half to two-thirds but are missing at the base. These shrubs are at their best in mid- to late spring, when their branch tips are smothered in white flowers reminiscent of apple blossoms.

Mountain Mahogany (*Cercocarpus betuloides*). P. 297.

Mountain mahogany has the hard wood of the tropical mahoganies used so widely for veneer, but is not related to them. Instead of being a large tree, it's a slender large shrub or small tree that occurs periodically through hard chaparral. Sometimes leafless in winter, the plant is then identified by its smooth grayish silver bark and flexible twigs. The leaves are beautiful up close, for their edges are serrated most of the way around, and the featherlike veins look like they've been imprinted into the leaf itself. When flowers are produced in abundance they may draw attention, but they seldom really distinguish themselves, for they lack the petals of most rosaceous flowers, having only greenish yellow sepals and pale yellow stamens. The show comes later toward summer as the single-seeded fruits carry long styles lined with plumelike silvery white hairs. When these are backlit the shrubs seem to glow. The plumed fruits represent an adaptation for dispersing the fruits on strong winds.

Toyon or California Holly (*Heteromeles arbutifolia*). P. 298.

Formerly known by the botanical name *Photinia*, toyon is a cousin but not a brother to the Chinese photinias so commonly grown as ornamental shrubs. The latter have bright red new leaves, whereas toyon does not, and toyon is an altogether stouter plant, attaining true tree status in coastal canyons of southern California. There, old toyons are shaped much like coast

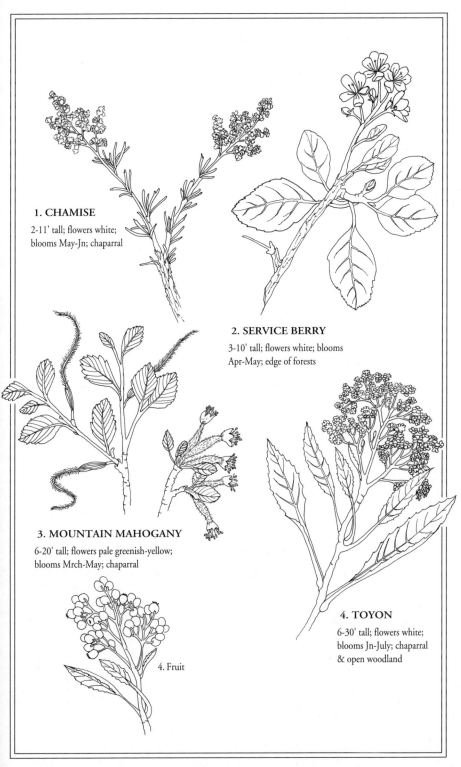

1. CHAMISE

2-11' tall; flowers white;
blooms May-Jn; chaparral

2. SERVICE BERRY

3-10' tall; flowers white; blooms
Apr-May; edge of forests

3. MOUNTAIN MAHOGANY

6-20' tall; flowers pale greenish-yellow;
blooms Mrch-May; chaparral

4. Fruit

4. TOYON

6-30' tall; flowers white;
blooms Jn-July; chaparral
& open woodland

live oaks, with rounded canopies, and they may occur in dense stands, creating a woodland all their own. Hollywood was named for just such a grove. The California holly there resembled a woodland of English true holly, which belongs to the unrelated family Aquifoliaceae. Toyon is a handsome shrub for the garden and has many endearing features: leathery hollylike leaves (but the serrated margins are not so spiny as those of true holly), large rounded clusters of white flowers in late spring, and vivid orange-red berries in late fall (making toyon an excellent holly substitute around the holidays). The species name "arbutifolia" indicates the likeness of the leaves to those of madrone; in fact, the leaves look very similar indeed, even though madrone belongs to the heather family, Ericaceae.

Creambush or Ocean Spray (*Holodiscus discolor*). P. 298.

The soft, fuzzy leaves of creambush are similar in toothing, vein pattern, and overall shape to those of mountain mahogany, but once you crush and smell them, they are immediately recognizable by their delightful fruity aroma. Winter deciduous, creambush wears a new mantle of pale green leaves in spring, then bursts into glorious blossom at spring's end. The flowers— tiny individually—are massed into dense creamy frothlike clusters reminiscent of their relatives the spiraeas. The old flowers hang on, turning rusty brown as they age, and may even remain on the shrub through the winter, a sure identifying characteristic when creambush is leafless and lifeless. The second common name—ocean spray—applies because this shrub is particularly abundant overlooking the ocean along the edge of forest knolls; in our area, look for it in protected, shady canyons.

Oso Berry (*Oemleria cerasiformis*). P. 298.

This small shrub is noted for its beauty in earliest spring. Living near canyon bottoms in wooded places, the delicate, pale green leaves emerge just as the drooping flower clusters open their chains of white, applelike blossoms. At the end of spring, flowers are replaced by small oval, purple fruits with a single seed or stone inside. Such fruits—related to the plums, cherries, apricots, and peaches of commerce—are referred to as drupes. Look for oso berry in habitats similar to those of the rare western leatherwood (*Dirca occidentalis*); out of flower the two shrubs look closely similar in size, shape, and leaves, but leatherwood's branches can be twisted into a circle without breaking.

Ninebark (*Physocarpus capitatus*). P. 298.

Lining vigorous brooks and streams, often in company with creek dogwood and wild roses, ninebark is a pretty, dense deciduous shrub of forest edges. Barren in winter, the twigs then seem to point every which way. The pale green new leaves in spring are interesting for their superficial resemblance to currant or gooseberry leaves, but close inspection shows that the lobes are actually pinnate and not palmate (as they are in currants). Dense, globe-shaped clusters of white flowers—individually like miniature white single roses—appear toward the end of spring, but the seed pods that follow are even more striking. Green when young, these pods ripen to a glossy red before finally fading brown and splitting into five-pointed stars. The inflated appearance of these seed pods is the reason for the genus name "physocarpus," meaning bladder fruit. Closer to the coast look for ninebark with bayberry and western azalea, the latter missing from our area. The common name alludes to the several layers of peely, brownish bark exposed on winter twigs.

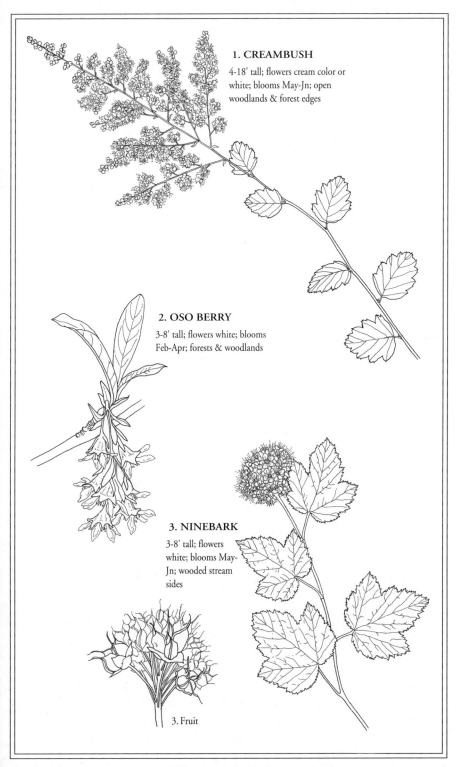

1. CREAMBUSH

4-18' tall; flowers cream color or white; blooms May-Jn; open woodlands & forest edges

2. OSO BERRY

3-8' tall; flowers white; blooms Feb-Apr; forests & woodlands

3. NINEBARK

3-8' tall; flowers white; blooms May-Jn; wooded stream sides

3. Fruit

Wild Plums and Cherries (*Prunus* spp.). P. 298.

To really see the wild plums and cherries at their best, you need to visit the Sierra or our northern mountains; the same species occur in our area but are rare and occur as isolated individuals. Two species are called cherries: *P. emarginata* (the bitter cherry) and *P. virginiana demissa* (choke-cherry). Bitter cherry creates small hedges or colonies where it grows, and spreads by creeping suckers. On a hot day the bitterness pervades not only the bright translucent red fruits but the silvery bark, which has a decidedly medicinal quality to its odor. The leaves are narrowly wedge shaped and like our other two prunuses are winter deciduous. Choke-cherry is a large shrub that occurs in slightly altered forms all across the United States. Its elliptical serrated leaves are larger and the bark less resinous than bitter cherry, and the flowers and fruits instead of being scattered in small clusters are aggregated into tight spikes. The deep red-purple fruits are intensely bitter when ripe but make excellent jams and jellies. Finally, our third prunus—*P. subcordata* (Sierra plum)—has decidedly plumlike leaves and small but good-tasting deep red-purple plumlike fruits. The flowers range from white to pinkish, appearing in small clusters just as leaves emerge in mid-spring. The species name subcordata alludes to the shallowly heart-shaped base of the leaves.

Wild Roses (*Rosa* spp.). P. 298.

Wild roses are easily recognized as close relatives to our garden roses, although the flowers are never double; rather, they resemble the old-fashioned rambling roses you may still see around early farmhouses in California. Distinguishing features of roses include the prickle-lined stems—spiny structures that occur between leaves—and also the pinnately compound leaves with fused stipules at their base, and the single roselike flowers, with the wonderful aroma of roses. Our species all have pinkish to lilac-mauve petals and large clusters of bright yellow stamens in the center. Flowers ripen into bright orange rose "hips," in reality not fruits at all but the fleshy sepal cups that surround the true fruits. Split a hip open; you'll see the several fuzzy, single-seeded achenes inside. Often the hips are the last bit of color for the year, as leaves are discarded in the fall in anticipation of winter. Our two roses are *Rosa californica* (California wild rose), a hedge-forming, stream-following species with flowers two to three inches across; and *R. gymnocarpa* (woodrose), a small, spreading shrub with finer prickles and miniature flowers of deep pink. The latter prefers moist forests and has sepalless hips, the reason for the name "gymnocarpa," which literally means naked fruit.

Thimbleberry (*Rubus parviflorus*). P. 298.

Most rubuses are prickly plants, such as the raspberries and blackberries, but thimbleberry differs by lacking spines altogether. The stems grow in multiple bunches, increasing their territory every year by underground roots. In winter, the leafless stems are characterized by long strips of semidetached brown bark. In spring, leaves reappear and look like a soft, fuzzy version of maple leaves. By mid-spring, the bushes are sprinkled with large white, single-roselike flowers that bely the species name "parviflorus," which literally means small flower. Perhaps the person naming this plant saw the blossoms dried on an herbarium sheet and so gave it this name, but actually the flowers are considerably larger than any on our other rubuses. By mid-summer, the thimble-shaped berries have ripened and turned dark red. At that time they're easily removed from the cone-shaped receptacle underneath (just as with raspberries) and are delectable, with a delicate flavor all their own.

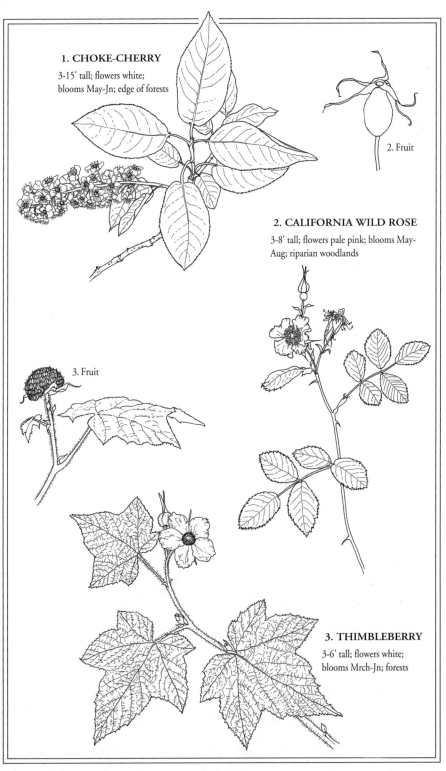

1. CHOKE-CHERRY

3-15' tall; flowers white;
blooms May-Jn; edge of forests

2. Fruit

2. CALIFORNIA WILD ROSE

3-8' tall; flowers pale pink; blooms May-
Aug; riparian woodlands

3. Fruit

3. THIMBLEBERRY

3-6' tall; flowers white;
blooms Mrch-Jn; forests

RUE OR CITRUS FAMILY (RUTACEAE).

Woody shrubs or trees; leaves with dark glands giving a bitter scent. Flowers in close clusters; 4 or 5 separate sepals, 4 or 5 separate petals attached to a nectar disc, same number or twice as many stamens, single pistil with superior ovary. Fruit a winged samara or fleshy "hesperidium."

Hopbush (*Ptelea crenulata*). P. 298.

The name hopbush refers to the fact that the bitter fruits were sometimes used as a substitute for hops in flavoring brews. The fruits that appear in mid-fall are single-seeded achenes almost completely surrounded by a thin, circular, whitish wing. The wings serve to carry seeds on winds. The fruits are ripe by the time the leaves have been shed, leaving only the new reddish-purple twigs sprinkled with white lenticels as a means of identification. Common in canyon bottoms around Mt. Diablo, hopbush requires a high water table during active growth. In spring, the compound leaves—consisting of three broad leaflets—reappear, and they also have a bitter odor. By late spring, the modest clusters of small white flowers have come and gone; up close, their shape and design is reminiscent of their relatives the citruses, although the flowers lack the sweet perfume of citrus blossoms.

SUMAC FAMILY (ANACARDIACEAE).

Shrubs, often with leaves divided into 3 parts. Small flowers in racemes; 5 sepals, 5 separate petals, 5 stamens by nectar disc, single pistil with superior ovary. Fruit a drupe.

Squawbush or Sourberry (*Rhus trilobata*). P. 298.

When people first spot squawbush, they think it's some kind of poison oak—indeed the two are closely related—but squawbush is really very different when you inspect it closely. Harmless to humans, its branches arch out and over (rather than being straight, as with poison oak) and have bark with a strong medicinal odor on hot days. The leaves are divided into three, but the shape of each leaflet is different from that in poison oak, and the leaves are dull and fuzzy, not smooth and shiny. In bloom, squawbush has tight racemes of outwardly directed pale yellow flowers, and by early fall, fuzzy red drupes that can be dried or made directly into a lemonadelike drink. Native Americans used the branches extensively in their basketry since they are very pliable.

Poison Oak (*Toxicodendron diversilobum*). P. 298.

Poison oak has earned for itself an indelible reputation; approach it always with care. Even those who have an immunity may later lose it; better to be safe than sorry. Should you touch the twigs (even the bare twigs in winter) or leaves, rinse your hands immediately and wash with mild soap: this should remove oils. Remember to wash your dog if it has accompanied you, and wash your clothes well. Despite the consequences of its toxicity poison oak is an interesting—even attractive—deciduous shrub. Thriving on disturbance, it has ventured into a wide range of different plant communities, behaving in each one according to circumstances: near the coast it lies prostrate next to the ground; in bright light it forms a dense shrub; in shade it climbs toward the sun, often ascending thirty to forty feet up a tree. New growth is signaled in early spring by a flush of glossy reddish new leaves. By mid-spring, dangling chains of whitish-green flowers perfume the air, attracting bees for pollination. This perfume is harmless to humans and actually enhances many a spring outing. By summer, there are whitish berries the birds consume, and in fall the foliage turns brilliant red before falling. When leaves color up in summer, it's a sign that the dry period has stressed the shrubs, telling them it's time to lose

1. Fruit

1. HOPBUSH

6-15' tall; flowers
white; blooms Apr-
May; canyons &
shaded slopes in
chaparral

2. SQUAWBUSH

2 ½-5' tall; flowers pale yellow;
blooms Mrch-Apr; canyon
bottoms in woodlands

2. Flowers

3. POISON OAK

3-10' tall (may climb higher);
flowers whitish green; blooms
Apr-May; many habitats

3. Fruit

leaves before they lose too much water. Out of leaf, poison oak may be identified by the long upright main branches with short, stubby side branches.

WATERLEAF FAMILY (HYDROPHYLLACEAE).

Occasional small shrubs, leaves often wool-covered beneath. Flowers in bud in coiled fiddleheads that unroll as flowers open; 5 partly fused sepals, 5 partly fused petals, 5 stamens attached to petals, single pistil with superior, 2-chambered ovary. Fruit a capsule.

Yerba Santa (*Eriodictyon californicum*). P. 297.

Yerba santa is the Spanish name for this shrub—meaning holy or sacred herb—since it has long been used medicinally. Because it belongs to the waterleaf family, yerba santa is unusual for its shrubby character; in fact, these are small opportunistic shrubs that grow along the periphery of chaparral or move in after burns. Once established, the running roots send up many new shoots to form large colonies. The leaves are distinctive and beautiful at close range (although their upper surface is sometimes disfigured by a black mold): the upper side is shiny, dark green and varnished on hot days; the lower side has a fine meshlike network of veins with clumps of tiny wooly hairs in between ("Eriodictyon" means wooly net). Yerba santa's leaves are pleasantly mint scented. It's claimed that chewing on one helps relieve thirst. By spring's end, many flowers are unfurling, with their attractive white to pale purple petals and nectar tubes, and butterflies eagerly seek them out for their nectar. In southern California, other species of yerba santa have leaves completely covered in dense, white-gray wool.

Lower surface of leaf

YERBA SANTA

2-7' tall; flowers pale purple or white;
blooms May-July; edge of chaparral &
after burns

Chapter 6
Vines of the East Bay

Cross-references to Vine Names

VINE FAMILIES

Scientific name	Common name	Page number
Aristolochiaceae	Birthwort family	120
Caprifoliaceae	Honeysuckle family	122
Convolvulaceae	Morning glory family	124
Cucurbitaceae	Gourd or cucumber family	122
Fabaceae	Pea family	124
Ranunculaceae	Buttercup family	120
Rosaceae	Rose family	126
Vitaceae	Vine family	128

VINE GENERA AND SPECIES

Common name	Scientific name	Page number
Berry, black	*Rubus* spp.	126
dew	*R. ursinus*	126
Himalayan	*R. procerus*	126
Bindweed	*Convolvulus arvensis*	124
Blackberry	*Rubus* spp.	126
Clematis	*Clematis* spp.	120, 122
Cucumber, Indian	*Marah* spp.	122
Grape, California wild	*Vitis californica*	128
Honeysuckle, vine	*Lonicera hispidula*	122
	and *interrupta*	122
Manroot	*Marah* spp.	122
Morning glories, wild	*Calystegia* spp.	124
Pipevine, California	*Aristolochia californica*	120
Sweet pea, wild	*Lathyrus* spp.	124, 126
Vetch	*Vicia* spp.	126
Virgin's bower	*Clematis* spp.	120, 122

Key to Vines

Flowers white or cream

1

- Tendrils; leaves varied, go to 3
- No tendrils, go to 2

2

- Stems not armed with spines, go to 2A
- Stems armed with stout spines, 5 leaflets per leaf
  Rubus procerus (Himalaya berry), p. 126
- Stems armed with slender prickles, 3 leaflets per leaf
  Rubus ursinus (dewberry), p. 126

2a

- Leaves compound into 3 or 5 leaflets
  Clematis spp. (virgin's bower), p. 120
- Leaves arrowhead shaped, simple, go to 2b
- Leaves elliptical, paired
  Lonicera spp. (vine honeysuckles), p. 122

2b

- Grows over shrubs or trees in natural areas
  Calystegia spp. (wild morning glory), p. 124
- Twines around plants or sprawls on ground in disturbed areas
  Convolvulus arvensis (bindweed), p. 124

3

- Leaves palmately lobed, flowers regular
  Marah spp. (Indian cucumber; manroot), p. 122
- Leaves pinnately compound, flowers pealike
  Lathyrus spp. (wild sweet pea), p. 124

Flowers brown/maroon

With green, pipe-shaped flowers
  Aristolochia californica (California pipevine), p. 120

Flowers greenish

1

- Tendrils; simple leaves
  Vitis californica (California wild grape), p. 126
- No tendrils; leaves divided into 3 leaflets
  Toxicodendron diversilobum (poison oak), p. 112

Flowers pink

1

- Stems armed with spines
  Rubus procerus (Himalaya berry), p. 126
- Stems not spiny, go to 2

2

- Saucer-shaped flowers; arrowhead-shaped leaves
  Convolvulus arvensis (bindweed), p. 124
- Two-lipped flowers; oval leaves
  Lonicera hispidula (vine honeysuckle), p. 122
- Pea-shaped flowers; pinnately compound leaves
  Lathyrus spp. (wild sweet pea), p. 124

Flowers purple to bluish

1

- Flowers fade tan or brown
  Lathyrus spp. (wild sweet pea), p. 124
- Flowers not fading these colors
  Vicia spp. (vetches), p. 126

Notes

Encyclopedia of Vines

BIRTHWORT FAMILY (ARISTOLOCHIACEAE).

Sometimes, deciduous vines with heart-shaped leaves. Flowers often irregular, in odd colors; 3 partly fused tepals, 9 stamens, single pistil with partly inferior ovary and 3 stigmas. Fruit a capsule.

California Pipevine (*Aristolochia californica*). P. 299.

California pipevine is our most curious vine and interesting just about any time of the year. Look for its flowers to be in bud at winter's end before the new leaves have opened. By early spring, vines are decorated with myriad pipes, giving rise to a second common name: Dutchman's pipe. Each pipe has an open mouth at its far end between three maroon-red tepal tips; these attractive to tiny flies that breed on rotting meat or other rotting substances. Once inside, they see the light streaming through the top of the pipe, for the upperside has translucent areas allowing light to pass through. Disoriented, the flies end up trapped at the stem end of the flower where stamens and stigmas are located. If you open a flower you'll release several of these tiny flies that apparently are trapped inside for some time. In fact, the first observers of this phenomenon believed that pipevine traps and digests insects, as do sundews or pitcher plants. Instead, as we now know, these flies are released eventually as the flower relaxes, so that they are free to visit and pollinate another flower. Virtually all true insectivorous plants use their leaves—not their flowers—for this purpose. After the pipes have done their job, myriad heart-shaped leaves appear. Later, at spring's end, pollinated flowers grow into nodding seed pods with six fluted angles around their periphery. Eventually they open to spill seeds far and wide. By late fall, the vines take a rest, losing their leaves through winter. One of the most interesting features of the pipevine is that it serves as the exclusive food for the larvae of the pipevine swallowtail butterfly—the adult a beautiful creature, with metallic black-purple wings.

BUTTERCUP FAMILY (RANUNCULACEAE).

Variable family, with only occasional vines, but leaves often divided into patterns of 3's. Flowers with separate sepals and petals (often only colored sepals and no petals); numerous, spirally arranged stamens; and several to many separate, distinct pistils. Fruit an achene or follicle.

Clematis or Virgin's Bower (*Clematis* spp.). P. 299.

The semiwoody stems of clematis scramble and clamber over any available shrubs or trees for support, sometimes appearing to nearly smother them with ropy stems. The compound leaves, borne in pairs, are divided into threes, fives, or even sevens, but those with three divisions superficially resemble leaves of poison oak, over which they sometimes grow. Clematis puts on a show twice: once in spring for its floral pageant of creamy starlike flowers and again in fall for its fruit clusters. Individual flowers are like four-pointed stars, filled with either myriad cream-color stamens or several greenish-yellow pistils according to sex. Watch these pistils as they develop into fruits; instead of the style falling away as the ovary expands to contain the developing seed, clematis retains and lengthens its styles. By the time fruits are ripe in fall, the styles have elongated into elegant, curved whitish plumes, which when massed are every bit as showy as the flowers. These plumes serve the purpose of carrying the single-seeded fruits on winds. Our two clematises look quite similar: *Clematis lasiantha* (pictured) has larger flowers in

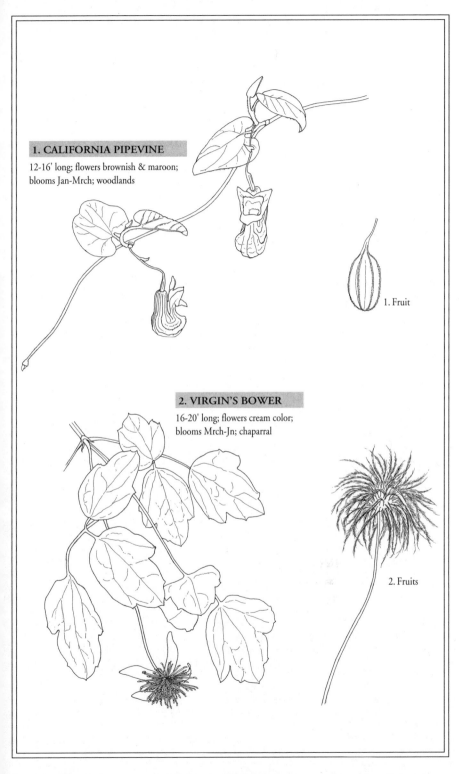

1. CALIFORNIA PIPEVINE

12-16' long; flowers brownish & maroon;
blooms Jan-Mrch; woodlands

1. Fruit

2. VIRGIN'S BOWER

16-20' long; flowers cream color;
blooms Mrch-Jn; chaparral

2. Fruits

mid-spring, and prefers the chaparral; *C. ligusticifolia* has smaller flowers in early summer and is a dweller in canyon bottoms, where it grows over riparian vegetation.

GOURD OR CUCUMBER FAMILY (CUCURBITACEAE).

Vines with palmately lobed leaves and curled tendrils. Flowers unisexual but borne on the same plant; 5 sepals, 5 partly fused petals, 5 fused stamens (for male flowers), single pistil with inferior ovary (for female flowers). Fruit a pepo, like a melon's.

Indian Cucumber or Manroot (*Marah* spp.). P. 299.

Although a member of the same family as the garden cucumber, Indian cucumber should never be mistaken for the vegetable, for the fruits are spiny and poisonous. In fact, the crushed seeds were among the items Native Americans used to stupefy fish. Manroot is perhaps a more appropriate name, for the mature plants go dormant in late fall to enormous, man-sized roots, which store quantities of food and water through the cold season. When spring days grow warm, the roots quickly send out several new shoots that grow with phenomenal speed to several feet long, often climbing and scrambling over other vegetation in their hurry to reach the sun. Soon, new leaves unfurl and side branches bear modest racemes of flowers. The first flowers are consistently male—stamens only—much as with their relatives, the squashes. Later flower clusters have one or more female flowers at the base of the cluster; these are easily recognized by the prickle-covered, green inferior ovary below the petals. The ovaries enlarge greatly and, by summer, have matured into curious-looking, spine-encrusted green "cucumbers." Be wary of the ripe fruits, for slight movement or touch will cause the pods to explode violently enough to forcibly eject the large seeds several inches. We have two manroots in our area: *Marah fabaceus* is most common, with smallish creamy starlike flowers that are open-bell shaped; the stouter *M. oreganus* has larger white flowers with the petals wide open.

HONEYSUCKLE FAMILY (CAPRIFOLIACEAE).

Sometimes woody, ropelike vines, with undivided, opposite leaves. Flowers (in these), irregular and 2-lipped, with 4 upper petals and 1 lower; 5 minute sepals, 5 partly fused petals, 5 stamens, and a single pistil with inferior ovary. Fruit (in these) a berry.

Vine Honeysuckle (*Lonicera hispidula vacillans*). P. 299.

The long, ropelike twisted stems of vine honeysuckle are familiar sights through our mixed-evergreen forests and oak woodlands. In winter, these are a bit eery—almost like something alien out of the tropics—but once in leaf they seem much more familiar. Starting life on the ground, the stems quickly find a shrub or tree to climb in order to get sufficient light to bloom. The paired, elliptical leaves are rather undistinguished, closely resembling those of snowberry, but once the flowers are about to appear, the last leaf pair is fused (becoming "perfoliate") around the stem just behind the flowers themselves. The modest white to rose-pink flowers are clearly irregular, with four upper and one lower petal curled back and with protruding yellow stamens. The flowers are not fragrant like many garden honeysuckles. The real show happens in early fall, when the fruits ripen to translucent red berries, beautiful to look at but with a frightful aftertaste. Two other honeysuckles from our area are worth noting: *L. interrupta* is a similar viny honeysuckle with pale yellow flowers and *L. subspicata denudata* is a curious half-shrub, half-vine, with whitish flowers.

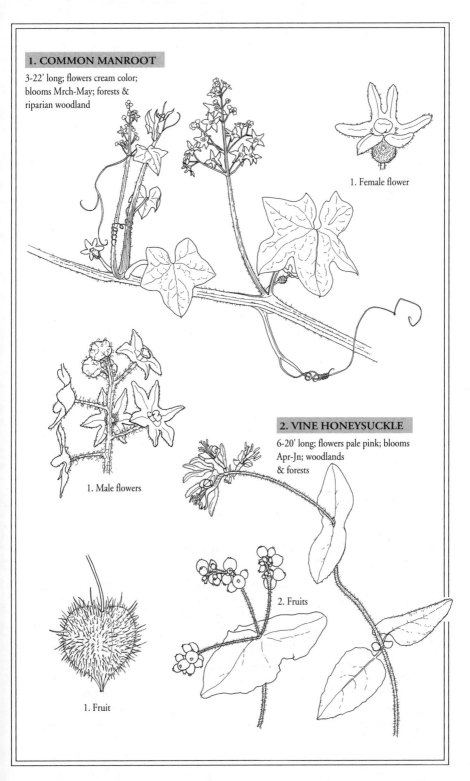

1. COMMON MANROOT

3-22' long; flowers cream color; blooms Mrch-May; forests & riparian woodland

1. Female flower

1. Male flowers

2. VINE HONEYSUCKLE

6-20' long; flowers pale pink; blooms Apr-Jn; woodlands & forests

2. Fruits

1. Fruit

MORNING GLORY FAMILY (CONVOLVULACEAE).

Mostly vines with twining stems and milky sap; leaves often arrowhead shaped. Flowers large and showy, funnel-shaped; 5 partly fused sepals, 5 very much fused petals pleated in bud, 5 stamens fused at base to petals, single pistil with 2-chambered, superior ovary. Fruit a capsule.

Wild Morning Glories (*Calystegia* spp.). P. 299, 300.

Our native morning glories have gotten a bad rap, partly because of their European cousin—bindweed—and partly because of the invasiveness of the garden morning glory (*Ipomoea purpurea*). Our species are all perennial vines from deep, woody roots and all die back each year to basal stems. The following spring they regrow to several feet in a matter of a few months. Although they do twine around native shrubs and trees, they seldom choke them or invade new territory. Early in the season, these vines are easily identified by their arrowhead-shaped leaves; later, the myriad white to pink- or purple-striped flowers transform the plants. By season's end, miniature woodroselike seed pods have formed, each with two or a few large, blackish seeds. In fact the resemblance to the Hawaiian woodrose is no accident, for that vine is also in the morning glory family: the roselike appearance is due to dried sepals, the center being the ovary or capsule itself. Our most common species is *Calystegia occidentalis*, the western morning glory. Two other species are worth mentioning. *C. subacaulis* forms low but broadly spreading mats of pale green leaves—the actual vine is buried underground—and the oversize cream-color blossoms sit among the leaves, elegant and decorative. A closely similar species—*C. malacophylla pedicellata*—has the same growth pattern but soft, wool-covered grayish leaves.

Bindweed (*Convolvulus arvensis*).

Bindweed is the bad guy of the morning glory world and comes to us from Europe. True to its name, the quickly growing vines wind themselves around anything handy, often literally smothering the host plant by their thick and rapid growth. Gardeners and farmers alike find controlling bindweed vexing, for the deeply seated roots grow horizontally and branch constantly, and discing or slicing them only creates more pieces. If you think that your bindweed has died out because it disappears in winter, think again; it's only taking a rest, and the new shoots emerge as soils warm in spring to start the whole process all over. It's a pity that bindweed is such a nuisance, for the inch-broad, saucer-shaped flowers are quite pretty, sometimes pure white, other times tinted pink.

PEA FAMILY (FABACEAE).

Nonwoody to woody plants, roots with nitrogen-fixing nodules, leaves usually pinnately compound or divided into 3s; stipules present. Flowers irregular and pea shaped, with one banner petal, two side wing petals, and 2 fused keel petals; 5 partly fused sepals; usually 10 stamens with 9 fused together by their filaments; single pistil with superior, 1-chambered ovary. Fruit a legume, like a pea pod.

Wild Sweet Peas (*Lathyrus* spp.). P. 299.

Sweet peas are native to Europe and California, but only the garden species (*L. odoratus*) is normally sweet scented. Most species are vines, climbing by tendrils: curled stalks produced at the ends of the pinnately compound leaves. One rank nonnative species—*L. latifolius*—adds large and colorful rose-pink blossoms in summer and is a common sight along shaded roadsides. Our most common native—*L. vestitus*, the decorated sweet pea—is common in woodlands and forest edges. Its pretty flowers are a mixture of white and pale purple with

1. WESTERN MORNING GLORY

4-20' long; flowers white, flushed pink-purple;
blooms Mrch-Jn; woodlands, chaparral

2. BINDWEED

1-4' long; flowers white
to pink; blooms May-
Sept; disturbed fields

3. DECORATED SWEET PEA

1-3' long; flowers pink-purple fading tan;
blooms Apr-May; woodlands & forests

3. Stigma and stamens

darker lines decorating the banner, but the old flowers fade to yellow-brown. *Lathyrus jepsonii,* Jepson's sweet pea, is a rare plant with pink-purple flowers and grows in brackish marshes in the Delta region.

Vetches (*Vicia* spp.). P. 299.

Vetches compose our other genus of scrambling to climbing viny peas, with pinnately compound leaves ending in curled tendrils. Most are introductions from Europe (for fodder), but a couple of species are also native. You can tell vetches apart from wild sweet peas (*Lathyrus*) by opening the flower and looking at the stigma. Vetches have a tiny tuft of hairs (use a hand lens) at the end of the style; sweet peas have a row of hairs along the lower side of the style tip (like a tooth brush). Two common introduced vetches are *V. sativa* (pictured), with single, pretty rose and purple flowers; and *V. villosa,* with one-sided racemes of deep blue-purple flowers. Native vetches include *V. americana,* a pretty sweet-pea-like perennial of coastal grasslands and open woodlands with short clusters of pink-purple flowers; and *V. gigantea,* giant vetch, which scrambles many feet over shrubs and bears small clusters of odd green-and-blue flowers. The latter is abundant on the edge of woodlands in the fog belt.

ROSE FAMILY (ROSACEAE).

Variable plants, with variable leaves and habit of growth. Flowers often showy; 5 sepals joined at base into a cup, 5 separate petals, usually numerous stamens. Pistils vary as to number and position of ovary. Fruits variable also.

Blackberries (*Rubus* spp.). P. 299.

Few families exhibit as much variability as the roses, and the rubuses themselves are variable; some spiny shrubs, others spineless shrubs, still others ground covers, and many vines. Our blackberries fit the last category, for even though they don't climb by tendrils or looped stems, their manner of growth is vinelike, the canes (stems) arching outward and downward at regular intervals and rooting where they touch ground, then repeating the process. In this manner, they soon comandeer considerable territory. On some the roots even send up new suckers to further the process. The white to pale pink blossoms appear in spring, like single roses, and are followed in summer by luscious near-black berries eagerly sought by berry pickers. The berries themselves really aren't berries botanically speaking: instead they're collections of several tiny, separate, fleshy pistils clumped together: each small unit is a separate pistil and the "stones" are the actual pits of these minute fruits. Our two blackberries differ in origin: *R. ursinus* (California dewberry) is a true native, with leaves divided into three pointed, spiny leaflets and with its canes covered with weak prickles. *R. procerus* [*R. discolor*] (Himalaya berry) comes from Asia and has leaves divided into five leaflets and stout, blood-letting spines on the canes. The latter is the one most often gathered for jams, jellies, and pies, since its fruits are considerably larger.

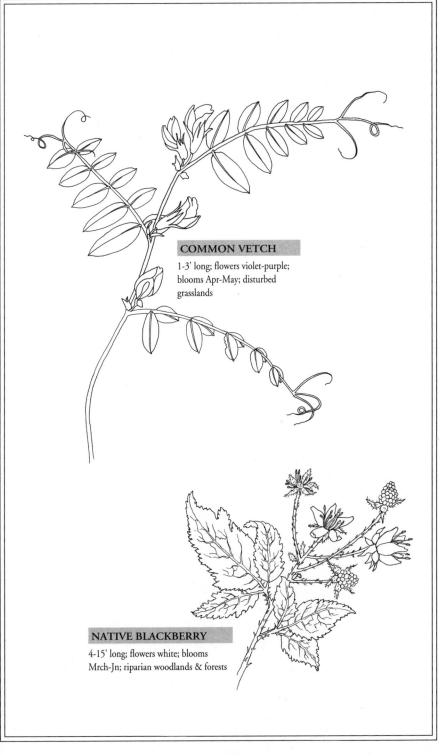

COMMON VETCH

1-3' long; flowers violet-purple;
blooms Apr-May; disturbed
grasslands

NATIVE BLACKBERRY

4-15' long; flowers white; blooms
Mrch-Jn; riparian woodlands & forests

Woody vines with broad, often palmately-lobed leaves and tendrils. Tiny greenish flowers in dense clusters; 5 minute sepals, no petals, 5 stamens, and single pistil with inferior ovary. Fruit a berry.

California Wild Grape (*Vitis californica*). P. 299.

Out of leaf and fruit, California wild grape creates great twisted woody ropes that climb high into the canopy of such riparian trees as cottonwoods, willows, alders, and maples. These are vines about as strong as any jungle lianas from which the legendary Tarzan swung. Even the cultivated grape would grow this way if given the chance; viticulturists trim and prune commercial vines so they'll be easy to handle and to force more energy into the yield of fruit. Although adapted to the same basic climate as the famed Old World wine grape (*V. vinifera*), eastern North American species are key to the whole wine and grape industry in California, for they provide the tough, resilient rootstock to which the fancy varieties are grafted. In its own right, California wild grape is a beauty: the myriad near-round bright green leaves decorate extravagant canopies in spring and summer. In fall they are heightened in two ways: by modest bunches of purple fruits, which are sour to the taste and seedy but great for jams or jellies, and by vibrant red, yellow, or purple leaves that turn color before falling.

CALIFORIA WILD GRAPE

6-50' long; flowers greenish; blooms
May-Jn; riparian woodland

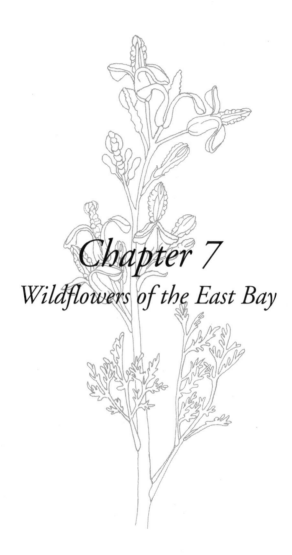

Chapter 7
Wildflowers of the East Bay

Cross-references
to Wildflower Names

	WILDFLOWER FAMILIES	
Common name	Scientific name	Page number
Parsley or umbel family	Apiaceae	246
Dogbane family	Apocynaceae	204
Aralia family	Araliaceae	176
Birthwort family	Aristolochiaceae	178
Milkweed family	Asclepiadaceae	230
Daisy or composite family	Asteraceae	188
Borage family	Boraginaceae	180
Mustard family	Brassicaceae	236
Bellflower family	Campanulaceae	176
Pink family	Caryophyllaceae	258
Rock-rose family	Cistaceae	266
Live forever family	Crassulaceae	228
Dodder family	Cuscutaceae	204
Spurge family	Euphorbiaceae	272
Pea family	Fabaceae	250
Fumitory family	Fumariaceae	216
Gentian family	Gentianaceae	218
Geranium family	Geraniaceae	218
Waterleaf family	Hydrophyllaceae	276
Iris family	Iridaceae	220
Mint family	Lamiaceae	232
Lily family	Liliaceae	220
Flax family	Linaceae	216
Blazing star family	Loasaceae	178
Mallow family	Malvaceae	230
Evening primrose family	Onagraceae	206
Orchid family	Orchidaceae	244
Broomrape family	Orobanchaceae	180
Poppy family	Papaveraceae	260
Plantain family	Plantaginaceae	260
Phlox family	Polemoniaceae	256
Buckwheat family	Polygonaceae	182
Portulaca family	Portulacaceae	264

Photo Gallery
Part II

Flame poppy only grows after fires have promoted seed germination. [photo Glenn Keator]

Wind poppy's seeds germinate well after fire. [photo Tim Dallas]

Golden eardrops, a bleeding heart relative, is also stimulated by fire. [photo Glenn Keator]

Fremont bush mallow is a fast-growing shrub that is a fire follower. [photo Glenn Keator]

California's grasslands turn green in winter and spring. [photo Leon Hunter]

Bunchgrasses, forming discrete clumps, were the original inhabitants of our grasslands. [photo Glenn Keator]

Later, grasslands turn brown as soils dry.
[photo Glenn Keator]

Grasslands still harbor rich variety of spring wildflowers.
[photo Tim Dallas]

Douglas's violet is an early spring perennial wildflower. [photo Tim Dallas]

Royal larkspur appears in mid-spring. [photo Glenn Keator]

Blow-wives exposes its fruits, encircled by flowerlike scales, in mid- to late spring. [photo Glenn Keator]

California poppy enjoys a long flowering season. [photo Tim Dallas]

Lupines create a prominent blue component in the floral pageant. [photo Glenn Keator]

Showy tarweed finishes the floral show in summer. [photo Glenn Keator]

Mariposa tulips open their exquisite butterflylike petals at spring's end. [photo Leon Hunter]

Forests show wind-pruned shapes of the evergreen crowns of California bay laurel. [photo Glenn Keator]

Bay's yellow flowers open in mid-winter when no other flower provides nectar. [photo Glenn Keator]

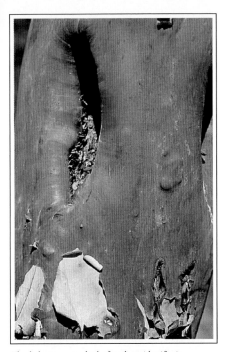

The sleek orange new bark of madrone identifies it.
[photo Glenn Keator]

Madrone flowers are nodding white urns filled with nectar.
[photo Glenn Keator]

California black oak's leaves turn yellow in fall.
[photo Glenn Keator]

California hazelnut, with dangling male catkins, is a
common forest treelet. [photo Glenn Keator]

California hazelnut's female flowers extend feathery red
stigmas to catch pollen. [photo Glenn Keator]

Pink-flowering currant has trusses of flowers in earliest spring. [photo Haskel Bazell]

The rare western leatherwood hangs its yellow bells before leaves reappear. [photo Haskel Bazell]

Canyon gooseberry hides tiny fuchsialike blossoms underneath spiny branches. [photo Glenn Keator]

An early forest wildflower is the deep red-flowered Indian warrior. [photo Glenn Keator]

The vivid pink fans of red ribbons clarkia appear toward spring's end. [photo Glenn Keator]

Redwoods of moderate age show a characteristic swooping out of their branches. [photo Glenn Keator]

Bacterial infections of dormant buds produce redwood burls. [photo Glenn Keator]

Redwood stump sprouts form a ring around the trunk base each year. [photo Glenn Keator]

Redwood bark is free of pitch and resists burning. [photo Glenn Keator]

The glossy evergreen leaves of huckleberry are common sights in redwood forests. [photo Glenn Keator]

Giant trillium's maroon flowers sit directly in the center of its three leaves. [photo Leon Hunter]

Coast wake-robin unfurls white flowers that fade deep rose, in early spring. [photo Glenn Keator]

Moist gullies are lined with spore-bearing plants, such as horsetails and ferns. [photo Glenn Keator]

Trillium's seed pod spills seeds with white oil bodies that entice ants for dispersal. [photo Glenn Keator]

Common name	Scientific name	Page number
Alfalfa	*Medicago sativa*	254
Alumroot, common	*Heuchera micrantha*	270
Angelica, wooly	*Angelica tomentosa*	246
Arnica, coastal	*Arnica discoidea*	190
rayless	*A. discoidea*	190
Aster	*Aster* spp.	190
broadleaf	*A. radulinus*	190
Baby-blue-eyes	*Nemophila menziesii*	276
Bedstraw	*Galium* spp.	228
climbing	*G. porrigens*	228
prickle-leaf	*G. andrewsii*	228
Bee plant	*Scrophularia californica*	214
Bellardia	*Bellardia trixago*	214
Bells, mission	*Fritillaria lanceolata*	224
whispering	*Emmenanthe penduliflora*	276
Bindweed	*Convolvulus arvensis*	124
Biscuit root	*Lomatium* spp.	248
Bitter root	*Lewisia rediviva*	266
Blazing star, common	*Mentzelia laevicaulis*	178
Lindley's	*M. lindleyi*	178
Bleeding heart, western	*Dicentra formosa*	218
Blow-wives	*Achyrachaena mollis*	188
Blue dicks	*Dichelostemma capitatum*	224
Blue sailors	*Cichorium intybus*	194
Brass buttons	*Cotula coronipifolia*	194
Brodiaea, blue dicks	*Dichelostemma capitatum*	224
harvest	*Brodiaea elegans*	222
Ithuriel's spear	*Triteleia laxa*	226
white	*T. hyacinthina*	226
Broomrape	*Orobanche* spp.	180
Buckwheat	*Eriogonum* spp.	182
naked	*E. nudum*	182
sulfur	*E. umbellatum*	184
Wright's	*E. wrightii*	182
Bur clover	*Medicago* spp.	254

Common Name	Scientific Name	Page Number
Buttercup, California	*Ranunculus californicus*	186
water	*R. aquatilis*	186
Butterwort, Brewer's	*Senecio breweri*	202
cobweb	*S. aronicoides*	202
Camas, death	*Zigadenus venenosus*	226
false	*Z.* spp.	226
Canchalagua	*Centaurium* spp.	218
Cardoon	*Cynara cardunculus*	192
Carrot, wild	*Daucus pusillus*	246
Caterpillar flower	*Phacelia* spp.	278
Cat's ear	*Hypochaeris* spp.	200
Chatterbox	*Epipactis gigantea*	244
Checker bloom	*Sidalcea malvaeflora*	230
Chicory	*Cichorium intybus*	194
Cinquefoil, sticky	*Potentilla glandulosa*	268
Clarkia, elegant	*Clarkia unguiculata*	206
red ribbons	*C. concinna*	206
winecup	*C. purpurea*	206
Clover, bur	*Medicago* spp.	254
cow	*Trifolium fucatum*	256
elk	*Aralia californica*	176
owl's	*Orthocarpus purpurascens*	212
red	*Trifolium pratense*	256
sweet	*Melilotus* spp.	254
tomcat	*Trifolium tridentatum*	256
white	*T. repens*	256
Columbine, red	*Aquilegia formosa*	184
Coral root, spotted	*Corallorhiza maculata*	244
Cream cups	*Platystemon californicus*	262
Cress, rock	*Arabis* spp.	236
water	*Nasturtium officinale*	240
winter	*Barbarea orthoceras*	238
Cudweed	*Gnaphalium* spp.	196
Daisy, common	*Erigeron philadelphicus*	194
Dandelion, mountain	*Agoseris* spp.	190
native	*A.* spp. and *Microseris* spp.	190
weedy	*Taraxacum officinale*	200
Deer weed	*Lotus scoparius*	252
Dock	*Rumex* spp.	184
Dodder	*Cuscuta* spp.	204
Downingia	*Downingia* spp.	176

Common Name	Scientific Name	Page Number
Hawkweed, white	*Hieracium albiflorum*	198
Hemlock, poison	*Conium maculatum*	246
water	*Cicuta douglasii*	246
Hemp, Indian	*Apocynum cannabinum*	204
Hemp, Indian	*Psoralea macrostachya*	254
Herald-of-summer	*Clarkia* spp.	206
Horkelia	*Horkelia* spp.	268
Hound's tongue	*Cynoglossum grande*	180
Indian paintbrush	*Castilleja* spp.	210
Iris, coast	*Iris longipetala*	303
Douglas	*I. douglasiana*	220
Ithuriel's spear	*Triteleia laxa*	226
Jewel flower, common	*Streptanthus glandulosus*	240
Mt. Diablo	*S. hispidus*	240
Jimson weed	*Datura meteloides*	242
Johnny-jump-up	*Viola pedunculata*	274
Jupiter's beard	*Centranthus ruber*	272
Lacepod	*Thysanocarpus* spp.	240
Larkspur, blue	*Delphinium* spp.	186
California	*D. californicum*	186
red	*D. nudicaule*	186
royal	*D. variegatum*	186
western	*D. hesperium*	186
Lettuce, miner's	*Montia perfoliata*	266
wild	*Lactuca* spp.	200
Lily, checker	*Fritillaria lanceolata*	224
mariposa	*Calochortus* spp.	222
Locoweed, Gambel's	*Astragalus gambellianus*	250
Looking glass plant	*Githopsis specularioides*	178
Lotus	*Lotus* spp.	252
deer broom	*L. scoparius*	252
Lupine, broadleaf	*Lupinus latifolius*	252
dove	*L. bicolor*	252
sky	*Lupinus nanus*	252
whorled	*L. densiflorus*	252
woodland	*L. formosus*	252
Madder, field	*Sherardia arvensis*	230
Mallow	*Malva* spp.	230
checker	*Sidalcea malvaeflora*	230

Common Name	Scientific Name	Page Number
Sanicle, poison	*Sanicula bipinnata*	250
purple	*S. bipinnatifida*	250
rock	*S. saxatilis*	250
tuberous	*S. tuberosa*	250
woodland	*S. crassicaulis*	250
Saxifrage, California	*Saxifraga californica*	270
Scarlet pimpernel	*Anagallis arvensis*	264
Self-heal	*Prunella vulgaris*	234
Shepherd's purse	*Capsella bursa-pastoris*	238
Shooting stars		
Henderson's	*Dodecatheon hendersonii*	264
Padre	*D. clevelandii patulum*	264
Skullcap, California	*Scutellaria californica*	236
tuberous	*S. tuberosa*	236
Skunk weed	*Navarretia* spp.	258
Snapdragon, wild	*Antirrhinum vexillo-calyculatum*	210
Soap plant	*Chlorogalum pomeridianum*	224
Solomon's seal, false	*Smilacina* spp.	226
fat false	*S. racemosa*	226
starry false	*S. stellata*	226
Sorrel, sheep	*Rumex acetosella*	184
Speedwell	*Veronica americana*	216
Spine flower	*Chorizanthe membrancea*	182
Spring beauty	*Montia gypsophiloides*	266
Spurge	*Euphorbia* spp.	272
native	*E. fendleri*	272
petty	*E. peplus*	272
spotted	*E. maculata*	272
Spurrey, common	*Spergula arvensis*	260
sand	*Spergularia* spp.	260
Star flower	*Trientalis latifolia*	264
Star lily, Fremont	*Zigadenus fremontii*	226
Star tulip, Oakland	*Calochortus umbellatus*	224
Stonecrop, common	*Sedum spathulifolium*	228
Strawberry, California	*Fragaria vesca*	268
woodland	*F. vesca*	268
Suncups	*Camissonia* spp.	206
Sunflower, California	*Helianthus californicus*	198
common	*H. annuus*	198
Mt. Diablo	*Helianthella castanea*	198

Key to Wildflowers

Introductory note: "Wildflowers" are herbaceous plants lacking wood or bark. Some wildflowers in the following key are included here for the sake of completeness but are not listed in the Encyclopedia. Each species that is in the Encyclopedia is so noted here by a page number after the name. The lack of a page number indicates that the species is not detailed any further.

You can start at the first step to arrive at basic flower groups or, if you're familiar with the groups, turn directly to them (p. 150). Groups are arranged by form of flower and also by flower color. Colors, which are repeated under each group, are generally as follows:
- Yellow
- Red or orange
- Rose-purple or pink
- Blue or purple
- White
- Greenish or brownish

Key to Groups

1
- Flowers arranged in daisylike heads, go to 2
- Flowers not arranged this way, go to 3

2
- Heads with both ray and disc flowers, Group XI
- Heads with ray flowers only, Group XII
- Heads with disc flowers only, Group XIII

3
- Sepals and petals in 3s and leaves net-veined, Group I
- Sepals and petals in 3s but leaves parallel-veined, Group II
- Sepals and petals in 2s, 4s or 5s, go to 4
- No sepals or petals, Group XIV

4
- Flowers irregular, go to 5
- Flowers regular, go to 6

5
- Flowers clearly two-lipped, with throat or tube between, Group III
- Flowers pealike, with banner, 2 wings, and keel, Group IV
- Flowers irregular but not pealike or two-lipped, Group V

6
- Petals separate, or joined just at base but not forming a ring or tube, go to 7
- Petals joined to form a tube or ring, Group VI

7
- Four crosslike petals, as in mustards, Group VII
- Petals not crosslike, but flowers small and arranged in umbels of umbels, Group VIII
- Petals not crosslike, and flowers not arranged in compound umbels, go to 8

8

- Stamens numerous, Group IX
- Stamens 10 or fewer, Group X

Keys to Genera within Groups

Group I. Flower Parts in 3s; Leaves Net-veined.

Because of the few entries, these plants are keyed below without regard to basic color categories.

1

- Leaves ginger scented, broadly heart shaped
  Asarum caudatum (wild ginger), p. 178
- Leaves not scented, not heart shaped, go to 2

2

- Only three leaves per plant, go to 3
- Several leaves per plant, go to 4

3

- Single flower on stalk above leaves; flowers open white and age rosy pink
  Trillium ovatum (wake-robin), p. 226
- Single flower borne directly above leaves; flowers may be white, maroon, green, or pink but remain the same color from opening until fading
  Trillium chloropetalum (giant trillium), p. 226

4

- Leaves compound in coarse, fernlike segments
  Vancouveria planipetala (inside-out flower)
- Leaves simple, often in basal rosettes, go to 5

5

- Plants of wet habitats, in mud or even partially submerged, go to 6
- Plants in ordinary soil, not favoring wet places, go to 5a

5a

- Flower tepals ending in sharp spine
  Chorizanthe spp. (spine flower), p. 182
- Flower tepals not spiny
  Eriogonum spp. (wild buckwheats), p. 182, 184

6

- Leaves arrowhead shape
  Sagittaria latifolia (arrowhead or Indian potato)
- Leaves broadly elliptical, go to 7

7

- Ovaries covered with burs (use a hand lens)
  Echinodorus berteroi (no common name)
- Ovaries smooth
  Alisma spp. (water plantain)

Group II. Flower Parts in 3s; Leaves Parallel Veined.

- A. Flowers yellow or orange
- B. Flowers rose-purple or pink
- C. Flowers blue or purple
- D. Flowers white
- E. Flowers greenish or brownish

Group IIA. Flowers yellow or orange

1

- Sepals and petals alike; leaves irislike
  Sisyrinchium californicum (yellow-eyed grass), p. 220
- Sepals and petals different in shape and size; leaves long and grasslike, go to 2

2

- Flowers clear yellow, go to 3
- Flowers orange, with dark spots
  Lilium pardalinum (leopard lily)

- Flowers hanging globes
  Calochortus pulchellus (Mt. Diablo globe tulip) p. 222
- Flowers upright cups
  Calochortus luteus (yellow mariposa tulip), p. 222

Group IIB. Flowers rose-purple or pink

1

- Flowers in umbels; onion odor
  Allium spp. (wild onions), p. 220
- Flowers not in umbels; no onion odor, go to 2

2

- Leaves distinctly mottled
  Scoliopus bigelovii (fetid adder's tongue)
- Leaves without mottling, go to 3

3

- Flowers hanging globes
  Calochortus albus (white fairy lantern), p. 222
- Flowers upright cups
  Calochortus venustus and *splendens* (mariposa tulips), p. 222
- Flowers small starlike vases in dense panicles
  Clintonia andrewsiana (bead lily)

Group IIC. Flowers blue or purple

1

- Leaves irislike, go to 2
- Leaves not irislike, go to 3

2

- Petals and sepals alike
  Sisyrinchium bellum (blue-eyed grass), p. 220
- Petals upright; sepals droop, go to 2a

2a

- Leaves stiff and upright, bluish green; no well-developed flower tube below petals
  Iris longipetala (coast iris)
- Leaves often gracefully curved, shiny dark green; definite flower tube below petals
  Iris douglasiana (Douglas iris), p. 220

3

- Flowers in congested clusters close together
  Dichelostemma spp. (blue dicks and ookow), p. 224
- Flowers on conspicuous stalks, go to 4

4

- Leaves and flowers onion scented; no flower tube below petals
  Allium spp. (wild onions), p. 220
- Leaves and flowers not onion scented; no flower tube; flowers cup shaped
  Calochortus splendens (lilac mariposa tulip), p. 220
- Like the last but flowers shallow saucer shaped
  Calochortus umbellatus (Oakland star tulip, p. 224
- Leaves and flowers not onion scented; definite flower tube below petals, go to 5

5

- Six fertile stamens
  Triteleia laxa (Ithuriel's spear), p. 226
- Three fertile stamens
  Brodiaea elegans (harvest brodiaea), p. 222

Group IID. Flowers white

1

- Flowers irregular
  Piperia spp. (rein orchids), p. 244
- Flowers regular, go to 2

2

- Sepals and petals different in shape and size
  Calochortus venustus (mariposa tulip), p. 222
- Sepals and petals similar in shape, size, and color, go to 3

3

- Flowers nodding; plant not onion scented
  Fritillaria liliacea (white fritillary)
- Flowers nodding, plant onion scented
  Allium triquetrum (nonnative onion)
- Flowers upright, go to 4

4

- Flowers in umbels, go to 5
- Flowers in racemes or panicles, go to 4a

4a

- Leaves wavy, at base of plant
  Chlorogalum pomeridianum (soap plant), p. 224
- Leaves flat, borne along stems
  Smilacina spp. (false Solomon's seals), p. 226

5

- Plants onion scented
  Allium amplectens (white lava onion)
- Plants lack onion odor, go to 5a

5a

- Petals fused to form a shallow tube, go to 6
- Petals completely separate to base
  Muilla maritima (no common name)

6

- Flowers borne on stalks much less than 2 inches long
  Triteleia hyacinthina (white brodiaea), p. 226
- Flowers borne on stalks well over 2 inches long
  Triteleia peduncularis (long-rayed brodiaea)

Group IIE. Flowers greenish or brownish

1

- Flowers irregular, go to 2
- Flowers regular, go to 4

2

- Green leaves (but may wither by time of flowering), go to 3
- No green leaves
  Corallorhiza maculata (spotted coral root), p. 244

3

- Flowers tiny, spurred
  Piperia spp. (rein orchids), p. 244
- Flowers about an inch across, not spurred
  Epipactis gigantea (stream orchid or chatterbox), p. 244

4

- Flowers nodding, checkered
  Fritillaria lanceolata (mission bells or checker lily), p. 224
- Like the last but flowers malodorous at close range
  Fritillaria agrestis (stink bells)
- Flowers hanging, plain green, hidden under leaves
  Disporum hookeri (fairy bells), p. 224
- Flowers upright; leaves mottled
  Scoliopus bigelovii (foetid adder's tongue)

Group III. Flowers Irregular, Two-lipped.

    A. Flowers yellow
    B. Flowers red or orange
    C. Flowers rose-purple or pink
    D. Flowers blue or purple
    E. Flowers white

Group IIIA. Flowers yellow

1

- Parasitic, without green leaves
  Orobanche spp. (broomrape), p. 180
- Green leaves present, go to 2

2

- Upper lip with 2 obvious petals, lower lip spotted
  Mimulus guttatus (golden monkey-flower), p. 212
- Like the last but no spots on lower lip; leaves slimy to touch
  Mimulus moschatus (musk monkey-flower), p. 212
- Upper lip with petals folded over stamens, go to 3

3

- Leaves sticky, not finely divided
  Parentucellia viscosa (no common name)
- Leaves smooth, finely divided
  Orthocarpus erianthus and *lithospermoides* (cream sacs), p. 212

Group IIIB. Flowers red or orange

1

- Upper lip folded over stamens, go to 2
- Upper lip with 2 obvious petals; flowers scarlet, go to 1a
- Upper lip with 2 obvious petals forming an awning; flowers maroon-red
  Scrophularia californica (California bee plant), p. 214

1a

- Upper lip with petals flared forward; streamside plants
  Mimulus cardinalis (scarlet monkeyflower), p. 210
- Upper lip with petals turned up or back; plants of dry, gravelly places
  Penstemon centranthifolius (scarlet bugler)

2

- Leaves fernlike; flowers deep crimson-red
  Pedicularis densiflorus (Indian warrior), p. 214
- Leaves not at all fernlike; flowers scarlet to orange
  Castilleja spp. (Indian paintbrush), p. 210

Group IIIC. Flowers rose-purple or pink

1

- Leaves strongly scented, go to 2
- Leaves not scented, go to 3

2

- Upper lip sticking straight up, go to 2a
- Upper lip like an awning
  Scrophularia californica (California bee plant), p. 214
- Upper lip folded around stamens and like a long beak
  Orthocarpus purpurascens (owl's clover), p. 212

2a

- Upper lip like a pair of mickey mouse ears
  Mimulus douglasii (mickey mouse monkeyflower), p. 212
- Upper lip not this way
  Mimulus spp. (other annual monkey-flowers)

3

- Flowers no more than an inch long; lower lip spotted
  Stachys spp. (wood mint), p. 236
- Flowers well over an inch long; no spots on lip
  Salvia spathacea (hummingbird sage), p. 234

Group IIID. Flowers blue or purple

1

- Green leaves present, go to 2
- Parasitic, without green leaves, go to 1a

1a

- Flowers borne on slender stems or fat asparaguslike stalks but not densely arranged in conelike clusters
  Orobanche spp. (broomrapes), p. 180
- Flowers borne in dense conelike clusters
  Boschniakia strobilacea (cone flower)

2

- Ovary superior, go to 3
- Ovary inferior, like a long stalk
  Downingia spp. (no common name), p. 176

3

- Leaves strongly scented, go to 4
- Leaves not scented, go to 6

4

- Flowers in whorls; leaves not turpentine scented, go to 5
- Flowers not in obvious whorls; leaves turpentine scented
  Trichostema lanceolatum (turpentine or vinegar weed), p. 236

5

- Leaves strongly mint scented, simple
  Pogogyne spp. (vernal pool mint), p. 234
- Leaves sage scented, deeply lobed, go to 5a

5a

- Leaves spiny; flowers pale blue
  Salvia carduacea (thistle sage)
- Leaves not spiny; flowers deep or clear blue
  Salvia columbariae (chia), p. 234

6

- All five petals obvious, go to 7
- Only four petals obvious; middle lower petal hidden
  Collinsia spp. (Chinese houses), p. 210
- Only four petals obvious; upper two folded around stamens
  Bellardia trixago (no common name), p. 214

7

- Flowers snapdragonlike; go to 7a
- Stamens, one sterile and 4 fertile; perennial
  Penstemon heterophyllus (foothill penstemon), p. 214
- 4 stamens only; tiny annual
  Tonella tenella (no common name)
- 4 stamens only; creeping perennial
  Prunella vulgaris (self-heal), p. 234

7a

- 4 stamens; sepals clearly 5, go to 7b
- 4 stamens; sepals completely fused
  Scutellaria tuberosa (tuberous skullcap), p. 236

7b

- Flowers spurred
  Linaria canadensis (blue toadflax)
- Flowers lack spur
  Antirrhinum vexillo-calyculata (wild snapdragon), p. 210

Group IIIE. Flowers white

1

- Leaves strongly scented, go to 3
- Leaves not strongly scented, go to 2

2

- Flowers snapdragonlike
  Scutellaria californica (California skullcap), p. 236
- Flowers not snapdragonlike, go to 2a

2a

- Plants form prostrate ground cover
  Lippia nodiflora (no common name)
- Plants with upright branches, go to 2b

2b

- Flowers borne singly or in pairs; lower lip inflated like a pelican beak
  Cordylanthus spp. (pelican beak)
- Flowers borne in dense spikes; lower lip not pelican-beak-like
  Orthocarpus attenuatus (no common name), p. 212

3

- Plants upright, go to 4
- Creeping plant with tiny flowers in leaf axils
  Satureja douglasii (yerba buena), p. 234

4

- Flowers very small, less than ½ inch long; lower lip not spotted
  Marrubium vulgare (horehound), p. 232
- Flowers over ½ inch long; lower lip usually spotted
  Stachys spp. (wood mints), p. 236

Group IV. Flowers Irregular, Pealike

A. Flowers yellow
B. Flowers rose-purple or pink
C. Flowers blue or purple
D. Flowers white
E. Flowers greenish or brownish

Group IVA. Flowers yellow

1

- Flowers borne singly between leaf and stem, go to 2
- Flowers borne in other arrangements, go to 3

2

- Leaflets in 3s; fruits coiled up, go to 2a
- Leaflets more than 3 per leaf; fruits not spiny
  Lotus spp. (lotus, bird's trefoil), p. 252

2a

- Fruits lack spines or barbs
  Medicago lupulina (no common name)
- Fruits clearly with spiny barbs, go to 2b

2b

- Leaves spotted
  Medicago arabica (spotted medick)
- Leaves lack spots
  Medicago polymorpha (common bur clover), p. 254

3

- Flowers in spikelike clusters, go to 4
- Flowers in umbels on subshrubs
  Lotus scoparius (Deer broom or deer weed), p. 252

4

- Flower spikes slender; stipules much smaller than leaflets
  Melilotus indicus (yellow sweet clover), p. 254
- Flower spikes broadly tapered; stipules conspicuous and almost as large as leaflets
  Thermopsis macrophylla (false lupine)

Group IVB. Flowers rose-purple or pink

1

- Vines with tendrils, go to 2
- Upright plants; no tendrils, go to 3

2

- Flowers borne in leaf axils
  Vicia sativa (common vetch), p. 126
- Flowers borne in tight clusters
  Lathyrus spp. (wild sweet pea), p. 124

3
- Leaves simple; sepals highly irregular
  Polygala californica (California milkwort)
- Leaves compound; sepals slightly irregular,
  go to 4

4
- Flowers borne in heads; leaflets 3 per leaf
  Trifolium spp. (clovers), p. 254
- Flowers borne in short spikes or heads;
  leaves pinnately compound
  Astragalus gambellianus and *didymocarpus*
  (annual rattlepods), p. 250
- Flowers borne singly; leaflets mostly more
  than 3 per leaf
  Lotus spp. (lotuses), p. 252

Group IVC. Flowers blue or purple

1
- Leaves pinnately compound; tendrils,
  go to 2
- Leaves palmately compound, go to 1a
- Leaves divided into 3; no tendrils, go to 1b

1a
- Flowers in racemes or spikes
  Lupinus spp. (lupines), p. 252
- Flowers in rounded heads close to ground
  Psoralea californica (no common name),
  p. 254

1b
- Plants seldom over 3 feet tall; flowers deep
  blue-purple
  Medicago sativa (alfalfa), p. 254
- Plants often top over 5 or 6 feet; flowers
  pale purple
  Psoralea macrostachya (Indian hemp),
  p. 254

2
- Flowers fade brownish
  Lathyrus spp. (wild sweet pea), p. 124
- Flowers don't fade brownish
  Vicia spp. (vetch), p. 126

Group IVD. Flowers white

1
- Leaves palmately compound
  Lupinus densiflorus (whorled lupine),
  p. 252
- Leaves divided into 3
  Melilotus alba (white sweet clover),
  p. 254
- Leaves pinnately compound, go to 2

2
- Ovary covered with prickles or barbs (use a
  hand lens)
  Glycyrrhiza lepidota (licorice root)
- Ovary smooth, without prickles
  Astragalus oxyphysus (locoweed or
  rattlepod)

Group IVE. Flowers greenish or brownish

1
- Leaves pinnately compound and pealike
  Lotus crassifolius (thick-leafed lotus),
  p.252
- Leaves divided into 3; strong scented
  Psoralea physodes (California tea), p. 254

Group V. Flowers Irregular, but Not Two-lipped or Pealike
  A. Flowers yellow
  B. Flowers red or orange
  C. Flowers rose-purple or pink
  D. Flowers blue or purple
  E. Flowers white

Group VA. Flowers yellow

1

- Petals 4; flower heart shaped
  Dicentra chrysantha (golden eardrops), p. 216
- Petals 5; flower violetlike, go to 2
- Petals 5; flowers neither heart shaped nor violetlike; petals arranged in wheellike fashion
  Verbascum thapsus (common mullein), p. 214

2

- Leaves deeply slashed; live in grasslands
  Viola douglasii (Douglas's violet), p. 274
- Leaves deeply slashed; live in scrub or woodlands
  Viola sheltonii (Shelton's violet), p. 274
- Leaves unlobed, go to 3

3

- Leaves nearly round on creeping plants
  Viola sempervirens (redwood violet), p. 274
- Leaves narrower; flowering stems upright, go to 4

4

- No purple or brown color on backside of upper petals
  Viola glabella (smooth yellow violet), p. 274
- Dark purple or brown on backside of upper petals, go to 5

5

- Leaves narrow, not heart shaped at base
  Viola purpurea (pine or oak violet), p. 274
- Leaves broad and heart shaped at base
  Viola pedunculata (wild pansy), p. 274

Group VB. Flowers red or orange

One species: *Delphinium nudicaule* (scarlet larkspur), p. 186

Group VC. Flowers rose-purple or pink

1

- Flowers spurred, not heart shaped; annuals of natural habitats
  Plectritis spp. (native valerians), p. 274
- Like the last but perennials of rocky, disturbed slopes
  Centranthus ruber (red valerian), p. 272
- Flowers not spurred, heart shaped
  Dicentra formosa (western bleeding heart), p. 218

Group VD. Flowers blue or purple

1

- Leaves definitely scented, go to 2
- Leaves not scented, go to 4

2

- Flowers borne in whorls on short spikes or heads, go to 3
- Flowers borne in tall spikes, not whorled
  Verbena lasiotstachys (common vervain), p. 274

3

- Flowers in heads
  Monardella spp. (coyote mint), p. 232
- Flowers in spikes
  Mentha pulegium (European pennyroyal), p. 232

4

- Flowers spurred (look carefully), go to 5
- Flowers not spurred
  Veronica spp. (speedwell), p. 216

5

- Leaves heart shaped
  Viola adunca (blue violet), p. 274
- Leaves not heart shaped; deeply lobed in pattern of a bird's foot
  Delphinium spp. (larkspur), p. 186

Group VE. Flowers white

1

- Leaves quilted; strong mint odor
  Mentha spicata (spearmint), p. 232
- Leaves not quilted but smooth; no odor
  Viola ocellata (western heartsease)

Group VI. Flowers Regular, Petals Joined

A. Flowers yellow
B. Flowers red or orange
C. Flowers rose-purple or pink
D. Flowers blue or purple
E. Flowers white
F. Flowers greenish or brownish

Group VIA. Flowers yellow

1

- Ovary inferior, often long and narrow, go to 2
- Ovary superior, hidden inside flower tube, go to 32

2

- Flowers open at night and late in day or early morning; stigma cross shaped
  Oenothera elata hookeri (Hooker's evening primrose), p. 208
- Flowers open in middle of day; stigma ball shaped
  Camissonia spp. (suncups), p. 206

3

- Flowers pale yellow; leaves deeply lobed
  Emmenanthe penduliflora (whispering bells), p. 276
- Flowers bright yellow to orange; leaves not lobed, go to 3a

3a

- Flowers coiled in bud, yellow-orange
  Amsinckia spp. (fiddleneck), p. 180
- Flowers not coiled in bud, clear yellow, go to 3b

3b

- Leaves slimy to touch; perennial plants
  Mimulus moschatus (musk monkey-flower), p. 212
- Leaves smooth to touch; tiny annuals
  Cicendia quadrangularis (yellow gentian)

Group VIB. Flowers red or orange

One species: *Zauschneria californica* (hummingbird fuchsia), p. 208

Group VIC. Flowers rose-purple or pink

1

- Ovary inferior, often narrow and stemlike in appearance, go to 2
- Ovary superior, go to 4

2

- Flowers borne in cylindrical, spiny-bracted spikes
  Dipsacus fullonum (Fuller's teasel)
- Flowers not in spikes; lack spiny bracts, go to 2a

2a

- Flowers less than ½ inch across; sepals not turned down, go to 3
- Flowers more than ½ inch across; sepals usually turned down
  Clarkia spp. (clarkia, farewell-to-spring; godetia), p. 206

3

- Leaves mostly fuzzy hairy; seeds lack hairs
  Boisduvalia spp. (no common name), p. 206
- Leaves usually smooth; seeds with tufts of white hairs
  Epilobium spp. (willow herb), p. 208

4

- Petals not swept back, go to 5
- Petals swept back
  Dodecatheon spp. (shooting stars), p. 264

5

- Two ovaries inside a cup, go to 6
- One ovary only, go to 7

6

- Flowers hang; petals threadlike
  Heuchera micrantha (common alum-root), p. 270
- Flowers horizontal; petals fringed
  Tellima grandiflora (fringe-cups), p. 270

7

- Stamens straight, go to 8
- Stamens clearly twisted
  Centaurium spp. (canchalagua), p. 218

8

- Leaves divided into fingerlike lobes
  Linanthus spp. (no common name), p. 256
- Leaves pinnately compound
  Polemonium carneum (foothill Jacob's ladder)
- Leaves simple, go to 9

9

- Petals notched
  Phlox gracilis (annual phlox), p. 258
- Petals not notched
  Collomia spp. (no common name)

Group VID. Flowers blue or purple

1

- Ovary inferior, go to 2
- Ovary superior, go to 3

2

- Upright flowers on tiny annual plants, go to 2a
- Nodding flowers on perennials up to a foot tall
  Campanula prenanthoides (California harebell), p. 176

2a

- Flowers upright bells
  Campanula exigua (annual harebell), p. 176
- Flowers upright saucers
  Githopsis specularioides (Venus looking glass), p. 178

3

- Flower buds clearly coiled up in fiddle-heads, go to 4
- Flower buds not coiled up, go to 7

4

- Style divided into two parts, go to 5
- Style not divided, go to 6

5

- Leaves with paler or yellowish spots
  Hydrophyllum occidentale (western waterleaf), p. 278
- Leaves lack spots, go to 5a

5a

- Stems with barbs that cling and are rough to touch
  Pholistoma auritum (fiesta flower), p. 278
- Stems lack barbs
  Phacelia spp. (caterpillar flower), p. 278

6

- Flowering stalks up to 2 or more feet tall; flowers over ½ inch across
  Cynoglossum grande (hound's tongue), p. 180
- Flowering stalks seldom over a foot tall; flowers less than ½ inch across
  Myosotis sylvaticus (forget-me-not)

7
- Leaves deeply lobed or compound, go to 8
- Leaves simple, sometimes toothed, go to 11

8
- Leaves pinnately lobed, go to 9
- Leaves divided like fingers on a hand
  Linanthus spp. (no common name), p. 256

9
- Flowers seldom more than ½ inch across; stigma split into 3 lobes, go to 10
- Flowers about an inch across; style 2-forked
  Nemophila menziesii (baby-blue-eyes), p. 276

10
- Flowers crowded into heads; without long spiny bracts
  Gilia capitata (globe gilia), p. 256
- Flowers crowded into heads, with long spiny bracts
  Navarretia spp. (skunkweed), p. 258
- Flowers not in heads, go to 10a

10a
- 3 colors present in flower
  Gilia tricolor (bird's eye gilia), p. 256
- Varied color combinations, but not 3 distinct colors
  Gilia spp. (other species than the two above)

11
- Papery flower tube extending down around ovary
  Lythrum spp. (loosestrife)
- No papery flower tube, go to 12

12
- Flowers funnel shaped, several inches long
  Datura spp. (Jimson weed), p. 242
- Flowers star shaped; less than an inch across
  Solanum nigrum (black nightshade), p. 244

Group VIE. Flowers white

1
- Not parasitic; green leaves present, go to 2
- Parasitic vine with orange stems
  Cuscuta spp. (dodder), p. 204

2
- Leaves with strong, unpleasant odor, go to 3
- Leaves not unpleasantly scented, go to 4

3
- Flowers flaring funnel shaped; stems not sticky
  Datura spp. (Jimson weed), p. 242
- Flowers narrow trumpets with petals turned out at ends; stems very sticky
  Nicotiana bigelovii (Indian tobacco), p. 242

4
- Flower buds rolled up in fiddlehead, go to 5
- Flower buds not rolled up, go to 8

5
- Plants mostly with stiff, white hairs, go to 6
- Plants lacking stiff, white hairs, go to 7

6
- Petals with tiny rim or disc at entrance to tube; style undivided, go to 6b
- Petals lack rim; style divided into 2 parts, go to 6a

6a
- Plants of wooded places; leaves sometimes spotted
  Hydrophyllum tenuipes (forest waterleaf)
- Plants of open places; leaves never spotted
  Phacelia spp. (caterpillar flower), p. 278

6b

- Ovary lobes (nutlets) held inside sepals after petals fall, not obvious
  Cryptantha and *Plagiobothrys* spp. (popcorn flowers), p. 180
- Ovary lobes (nutlets) spreading and hiding sepals; nutlets lined with comblike prickles
  Pectocarya spp. (comb-fruit)

7

- Tiny earlobelike green segments between sepals; leaves without special smell
  Nemophila spp. (woodland lover), p. 276
- No earlobelike segments; leaves fragrant (especially when dry)
  Eucrypta chrysanthemifolia (no common name), p. 276

8

- Leaves divided into fingerlike segments; flowers upright
  Linanthus spp. (no common name), p. 256
- Leaves simple but scalloped; flowers hang
  Heuchera micrantha (common alumroot), p. 270

Group VIF. Flowers greenish or brownish

1

- No milky juice, go to 2
- Abundant milky juice present in stems and leaves
  Apocynum cannabinum (Indian hemp), p. 204

2

- Plants lack special odor; flowers apparent, go to 3
- Plant ginger scented; flowers hide under leaves
  Asarum caudatum (wild ginger), p. 178

3

- Flowers in dense spikes; petals parchment-paper-like, go to 3a
- Flowers in racemes; petals fringed, not parchment-paper-like
  Tellima grandiflora (fringe-cups), p. 270

3a

- Leaves long and linear, densely hairy
  Plantago erecta (native plantain), p. 260
- Leaves lance shape, obscurely hairy
  Plantago lanceolata (English plantain)
- Leaves broadly oval, not hairy
  Plantago major (broadleaf plantain)

Group VII. Flowers Regular, 4 Petals in a Cross, Like Mustards

A. Flowers yellow
B. Flowers red or orange
C. Flowers rose-purple or pink
D. Flowers blue or purple
E. Flowers white
F. Flowers greenish or brownish

Group VIIA. Flowers yellow

1

- Grow in wet places; leaves smooth, go to 2
- Grow in dry places; leaves rough or hairy
  Brassica spp. (mustards), p. 236

2

- Leaves deeply divided into fernlike segments
  Descurainia spp. (tansy mustards)
- Leaves pinnately divided but not fernlike, go to 3

3

- Terminal leaflet of leaf much larger than other leaflets
  Barbarea orthoceras (winter cress), p. 238
- Leaflets all alike
  Rorippa curvisiliqua (yellow cress)

Group VIIB. Flowers red or orange

One species: *Erysimum capitatum* (foothill wallflower), p. 238

Group VIIC. Flowers red-purple or pink

1

- Flowers an inch or so across, on stems over a foot tall
  Raphanus sativa (radish weed), p. 240
- Flowers much less than an inch across, on stems generally much less than a foot tall
  Arabis spp. (rock cresses), p. 236

Group VIID. Flowers blue or purple

1

- Flowers an inch or so across, on stems over a foot tall
  Raphanus sativa (radish weed), p. 240
- Flowers much less than an inch across, on stems generally much less than a foot tall
  Streptanthus hispidus (Mt. Diablo jewel flower), p. 240

Group VIIE. Flowers white

1

- Grows on dry land, go to 2
- Grows in water
  Nasturtium officinale (water cress), p. 240

2

- Leaves mostly the same shape, go to 3
- Leaves of two entirely different shapes: round and simple or compound
  Dentaria californica (milkmaids), p. 238

3

- Seed pods as broad as long, go to 4
- Seed pods long and narrow, go to 3a

3a

- Petals crimped or wavy
  Thelypodium spp. (no common name)
- Petals not crimped or wavy, go to 3b

3b

- Flowers around an inch across
  Raphanus sativa (radish weed), p. 240
- Flowers much smaller, go to 3c

3c

- Leaves deeply pinnately lobed, mostly basal
  Cardamine oligosperma (bitter cress), p. 238
- Leaves not lobed, occur at base of plant and up stem
  Arabis glabra (tower mustard), p. 236

4

- Flowers in large flat-topped clusters
  Cardaria draba (white top)
- Flowers not clustered this way, go to 4a

4a

- Seed pods with definite circular crimped or fluted rim
  Thysanocarpus spp. (lacepod or fringepod), p. 240
- Seed pods not rimmed; triangular
  Capsella bursa-pastoris (shepherd's purse), p. 238
- Seed pods not rimmed; round, go to 4b

4b

- Seed pods rounded with notched tip
  Lepidium spp. (pepper-grass), p. 240
- Like the last but no notch at tip, go to 4c

4c

- Single-seeded seed pods
  Athysanus pusillus (no common name)
- More than one seed per pod
  Draba verna (annual draba)

Group VIIF. Flowers greenish or brownish

One species: *Streptanthus glandulosus* (common jewel flower), p. 240

Group VIII. Flowers Regular, Small, Arranged in Compound Umbels

A. Flowers yellow
B. Flowers red or orange
C. Flowers bluish
D. Flowers white
E. Flowers greenish or brownish

Group VIIIA. Flowers yellow

1

- Plants not licorice scented; seldom over 1 foot tall, go to 2
- Plants strongly licorice scented; several feet tall
  Foeniculum vulgare (fennel), p. 248

2

- Flower umbels very tight, almost buttonlike; foliage coarsely divided
  Sanicula spp. (sanicles), p. 250
- Flower umbels open; foliage finely divided, often fernlike
  Lomatium spp. (Indian biscuit root), p. 248

Group VIIIB. Flowers red or orange

One species: *Sanicula bipinnatifida* (purple sanicle), p. 250

Group VIIIC. Flowers bluish

One genus: *Eryngium* spp. (button parsleys)

Group VIIID. Flowers white

1

- Growing in water, go to 2
- Growing on dry land, go to 4

2

- Plants upright, often several feet tall; go to 3
- Plants creep and root as they grow, seldom over 1 foot tall
  Oenanthe sarmentosa (water parsley), p. 248

3

- Lower leaves twice divided; upper roots chambered
  Cicuta douglasii (water hemlock), p. 246
- Lower leaves once divided; upper roots not chambered
  Berula erecta (water parsnip), p. 246

4

- Leaves finely divided, sometimes fernlike, go to 5
- Leaves coarsely divided, never fernlike, go to 7

5

- Stems lacking purple spots, go to 6
- Stems and leaf petioles with purple spots and splotches
  Conium maculatum (poison hemlock), p. 246

6

- Fruits with long needlelike points; plants seldom over 1 foot tall
  Scandix pecten-veneris (shepherd's needles), p. 250
- Fruits lack needlelike points, go to 6a

6a

- Ovary with hooked bristles or barbs (use a hand lens), go to 6b
- Ovary with nonhooked bristles (use a hand lens), go to 6d
- Ovary lacks barbs or bristles
  Perideridia spp. (yampah), p. 250

6b

- No obvious hairs on plant
  Apiastrum angustifolium (false celery)
- Definite hairs on stems and leaves, go to 6c

6c

- Bristles in vertical rows following ovary ribs (use a hand lens)

  Caucalis microcarpa (no common name)

- Bristles all over surface of ovary (use a hand lens)

  Torilis spp. and *Anthriscus scandicina* (bur chervil)

6d

- Main umbels changing shape with age, basketlike when old

  Daucus carota (Queen Anne's lace or wild carrot), p. 246

- Main umbels not changing shape with age, more or less flat topped

  Daucus pusillus (native wild carrot), p. 246

7

- Main umbels often a foot or more across, go to 7a

- Main umbels seldom over 6 inches across

  Osmorhiza spp. (sweet cicely), p. 250

7a

- Leaves with few coarse divisions, ragged in age

  Heracleum lanatum (cow parsnip), p. 248

- Leaves with several divisions, tougher and seldom ragged in age

  Angelica tomentosa (wooly angelica), p. 246

Group VIIIE. Flowers greenish or brownish

1

- Small herbs with leaves less than a foot long, go to 2

- Giant herbs with leaves several feet long

  Aralia californica (elk clover or California aralia), p. 176

2

- Leaves round and scalloped

  Hydrocotyle spp. (marsh pennyworts)

- Leaves not rounded but much divided, go to 2a

2a

- Leaves with spine-tipped lobes; flowers in dense buttonlike umbellets

  Eryngium spp. (button parsley)

- Leaves lack spines; flowers in open umbellets, go to 2b

2b

- Plants celery scented

  Tauschia spp. (no common name)

- Plants anise scented

  Osmorhiza spp. (sweet cicely), p. 250

Group IX. Flowers Regular, Stamens Numerous

A. Flowers yellow

B. Flowers red or orange

C. Flowers rose-purple or pink

D. Flowers blue or purple

E. Flowers white

F. Flowers greenish or brownish

Group IXA. Flowers yellow

1

- Ovary superior, go to 2

- Ovary inferior, go to 1a

1a

- Flowers more than an inch across, go to 1b

- Flowers much less than an inch across

  Mentzelia dispersa and *micrantha* (blazing stars), p. 178

- Flowers clear deep yellow; petals broad
  Mentzelia lindleyi (Lindley's blazing star),
  p. 178
- Flowers pale yellow; petals narrow
  Mentzelia laevicaulis (common blazing
  star), p. 178

2

- Sepals remain on flower as petals open,
  go to 3
- Sepals fall as petals open
  Platystemon californicus (cream cups),
  p. 262

3

- Herbaceous plants; no rushlike stems,
  go to 4
- Subshrub with green, rushlike stems
  Helianthemum scoparium (rush-rose),
  p. 266

4

- Leaves simple, unlobed, opposite, go to 6
- Leaves lobed or compound, alternate,
  go to 5

5

- Leaves palmately lobed; lack stipules
  Ranunculus spp. (buttercups), p. 186
- Leaves pinnately compound; have stipules
  Potentilla glandulosa (sticky cinquefoil),
  p. 268

6

- Plants sprawling; flowers salmon-yellow
  Hypericum anagalloides (tinker's penny)
- Plants upright; flowers bright yellow
  Hypericum perforatum and *formosus*
  (Klamath weed and native St. Johnswort)

Group IXB. Flowers red or orange

1

- Upright flowers; no spurs, go to 2
- Hanging flowers with 5 spurs
  Aquilegia formosa (red columbine),
  p. 184

2

- Leaves pinnately lobed, but not fernlike;
  flowers deep red-orange, go to 3
- Leaves finely divided like ferns; flowers
  orange or yellow orange
  Eschscholzia spp. (California poppy),
  p. 262

3

- Petals green at base
  Papaver californicum (flame poppy),
  p. 262
- Petals dark purple at base
  Stylomecon heterophylla (wind poppy),
  p. 262

Group IXC. Flowers rose-purple or pink

1

- Leaves thin; stamens fused together by
  their filaments to form a tube, go to 2
- Leaves fleshy; stamens not fused
  Lewisia rediviva (bitterroot), p. 266

2

- Leaves all same shape; flowers less than ½
  inch across
  Malva spp. (mallows or cheeses), p. 230
- Leaves change shape from base of plant to
  top of stem; flowers at least an inch across
  Sidalcea malvaeflora (checker bloom),
  p. 230

Group IXD. Flowers blue or purple

One genus: *Malva* spp. (mallow or cheeses),
p. 230

Group IXE. Flowers white

1
- Sepals fall as petals open, go to 2
- Sepals remain on flower as petals open, go to 3

2
- Spiny plants with flowers several inches across
    Argemone munita (prickly poppy), p. 260
- Spineless plants with flowers seldom over an inch across, go to 2a

2a
- Petals cream color, often marked yellow; stamens flat
    Platystemon californicus (cream cups), p. 262
- Petals white; stamens not flattened
    Meconella californica (no common name)

3
- Leaves have definite stipules (be sure to find base of leaves), go to 4
- Leaves lack stipules, go to 5

4
- Leaves divided into several segments; fragrant
    Horkelia spp. (no common name), p. 268
- Leaves compound into 3 leaflets; no fragrance
    Fragaria vesca (woodland strawberry), p. 268

5
- Plants grow in water
    Ranunculus aquatilis (water buttercup), p. 186
- Plants grow on dry land
    Isopyrum spp. (rue-anemone), p. 186

Group IXF. Flowers greenish or brownish

1
- Conspicuous coarsely divided fernlike leaves and unspurred sepals
    Thalictrum fendleri polycarpum (foothill meadow rue), p. 200
- Simple leaves in low rosettes and long-spurred sepals
    Myosurus spp. (mousetail)

Group X. Flowers Regular, Stamens 10 or Fewer

A. Flowers yellow
B. Flowers red or orange
C. Flowers rose-purple or pink
D. Flowers blue or purple
E. Flowers white
F. Flowers greenish or brownish

Group XA. Flowers yellow

1
- Ovary inferior, go to 2
- Ovary superior, go to 3

2
- Flowers open at night or early or late in day; stigma cross shaped
    Oenothera elata hookeri (Hooker's evening primrose), p. 208
- Flowers open at midday; stigma ball shaped
    Camissonia spp. (suncups), p. 206

3
- Leaves divided into 3s, go to 3b
- Leaves simple, go to 3c

3b
- Flowers much less than an inch across; leaves often reddish
    Oxalis corniculata (weedy oxalis)
- Flowers an inch or more across; leaves never reddish
    Oxalis pes-caprae (Bermuda buttercup)

3c

- Leaves succulent and fleshy, go to 4
- Leaves thin, not fleshy, go to 3d

3d

- Flowers in dense clusters
  Eriogonum umbellatum (sulfur buckwheat), p. 184
- Flowers at tips of branches spaced apart
  Hesperolinon, (native flax), p. 216

4

- Annuals dying as soils dry out, go to 5
- Perennials, go to 6

5

- Five separate pistils per flower
  Parvisedum pentandrum (annual stonecrop)
- One pistil per flower
  Portulaca oleracea (purslane), p. 266

6

- Leaves broadly lance shaped; flowering stalk coming out between leaves
  Dudleya cymosa (hot rock dudleya), p. 228
- Leaves spoon shaped; flowering stalk coming out from center of leaves
  Sedum spathulifolium (common stonecrop), p. 228

Group XB. Flowers red or orange

1

- Petals not slashed, go to 2
- Petals conspicuously slashed or fringed
  Silene californica (Indian pink), p. 260

2

- Leaves thin, not fleshy, go to 3
- Leaves thick and fleshy
  Dudleya cymosa (hot rock dudleya), p. 228

3

- Orange flowers with dark purple center; leaves elliptical
  Anagalis arvensis (scarlet pimpernel), p. 264
- Red-brown flowers; leaves arrowhead shaped
  Rumex acetosella (sheep sorrel), p. 184

Group XC. Flowers rose-purple or pink

1

- Milky juice lacking, go to 2
- Milky juice obvious when stem or leaf broken
  Asclepias californica (California milkweed), p. 232

2

- Ovary inferior, go to 3
- Ovary superior, go to 5

3

- Flowers less than ½ inch across, sepals not turned down, go to 4
- Flowers more than an inch across, sepals usually turned down
  Clarkia spp. (clarkia, farewell-to-spring, godetia), p. 206

4

- Leaves fuzzy hairy; seeds smooth and hairless
  Boisduvalia spp. (no common name), p. 206
- Leaves smooth; seeds with tufted white hairs
  Epilobium spp. (willow herb), p. 208

5

- Leaves thick, fleshy, succulent, go to 6
- Leaves thin, not fleshy, go to 7

6

- Flowers red-purple
  Calandrinia ciliata (red maids), p. 264
- Flowers pale pink
  Montia spp. (miner's lettuce and relatives), p. 266

7

- Styles forming beaks on fruits, go to 8
- Styles shed with petals, not retained on fruits, go to 9

8

- Leaves pinnately lobed or divided
  Erodium spp. (filaree, clocks), p. 218
- Leaves palmately lobed
  Geranium spp. (wild geranium), p. 218

9

- Leaves divided into 3s
  Oxalis oregana (redwood sorrel)
- Leaves undivided, simple (may be lobed), go to 9a

9a

- Leaves basal or alternate, go to 10
- Leaves whorled, flowers starlike
  Trientalis latifolia (star flower), p. 264
- Leaves opposite, go to 9b

9b

- Petals notched; flowers pink to rose
  Silene verecunda (coast pink)
- Petals unnotched; flowers white to pale pink
  Silene gallica (windmill pinks), p. 258

10

- Petals and sepals distinct from one another
  Tellima grandiflora (fringe-cups), p. 270
- Perianth parts look alike
  Polygonum amphibium (water knotweed or smartweed)

Group XD. Flowers blue or purple

1

- Leaves basal or alternate, go to 2
- Leaves opposite; flowers palest purple
  Spergularia spp. (sand spurrey), p. 260
- Leaves whorled; flowers lavender-purple
  Sherardia arvensis (field madder), p. 230

2

- Leaves narrow, simple, unlobed
  Linum usitatissimum (common flax), p. 216
- Leaves broad, deeply lobed or divided
  Erodium spp. (filaree, clocks), p. 218

Group XE. Flowers white

1

- Leaves basal or alternate, go to 2
- Leaves opposite or whorled, go to 4

2

- Petals whole, not fringed, go to 3
- Petals fringed or slashed
  Lithophragma spp. (woodland star), p. 270

3

- Stems and leaves with abundant milky sap
  Euphorbia spp. (spurges), p. 272
- Stems and leaves lack milky juice, go to 3a

3a

- Flowers hang, bell shaped
  Heuchera micrantha (common alumroot), p. 270
- Flowers upright, not bell shaped, go to 3b

3b

- Leaves deeply pinnately divided; plants of vernal pools
  Limnanthes douglasii (common meadow foam)
- Leaves simple and unlobed; plants never in vernal pools, go to 3c

- Flowers 2 inches or more across, fading pink
    Oenothera deltoides howellii (Antioch dunes evening primrose), p. 208
- Flowers much less than an inch across, go to 3d

3d

- Leaves with papery stipules
    Polygonum spp. (knotweed, tearthumb)
- Leaves lack papery stipules, go to 3e

3e

- Leaves thick and fleshy, sprawling plant
    Calyptridium monandrum (no common name)
- Leaves thin, not fleshy, flowering stalks upright, go to 3f

3f

- Leaves narrow, almost linear
    Hesperolinon spp. (native flaxes), p. 216
- Leaves broad, spoon shaped or oblong, go to 3g

3g

- Leaves wooly covered beneath
    Eriogonum spp. (wild buckwheats), p. 182
- Leaves hairy but not wool covered
    Saxifraga californica (California saxifrage), p. 270

4

- Leaves whorled, go to 5
- Leaves opposite, go to 7

5

- Ovary superior; rounded stems, go to 6
- Ovary inferior; square stems
    Galium spp. (bedstraws), p. 228

6

- Leaves nearly linear, hairy
    Spergula arvensis (spurrey), p. 260
- Leaves broadly ovate, smooth
    Trientalis latifolia (star flower), p. 264

7

- Leaves thin, not fleshy; nodes swollen where leaves are attached, go to 8
- Leaves fleshy; nodes not swollen where leaves are attached
    Montia spp. (miner's lettuce and relatives), p. 266

8

- Sepals fused, covered with sticky hairs
    Silene gallica (windmill pinks), p. 258
- Sepals separate without sticky hairs, go to 9

9

- Petals notched to deeply slashed, go to 10
- Petals not notched or slashed, go to 9a

9a

- Leaves slender and grasslike
    Sagina occidentalis (western pearlwort)
- Leaves narrow lance shape, not at all grasslike
    Arenaria douglasii (Douglas's sandwort), p. 258

10

- Petals longer than sepals, deeply slashed
    Stellaria media (common chickweed)
- Petals longer than sepals, notched
    Arenaria macrophylla (perennial sandwort), p. 259
- Petals shorter than sepals (or missing), notched
    Cerastium spp. (mouse-eared chickweed)

Group XF. Flowers greenish or brownish

1
- Plants with milky juice when stems or leaves broken, go to 2
- No milky juice, go to 3

2
- Flowers in open umbels, easy-to-see parts
  Asclepias fascicularis (whorled milkweed), p. 232
- Flowers in tiny cups, appearing like single minute flowers all together
  Euphorbia spp. (spurges), p. 272

3
- Leaves thin; plants not minute, go to 4
- Leaves fleshy on minute plants
  Crassula erecta and *aquatica* (pygmy weed), p. 228

4
- Leaves not whorled; stems rounded, go to 5
- Whorled leaves on square stems
  Galium spp. (bedstraws), p. 228

5
- Leaves entire, never toothed, go to 6
- Leaves ovate, coarsely toothed; stems armed with stinging hairs
  Urtica spp. (stinging nettle), p. 242
- Leaves rounded and scalloped with teeth; no stinging hairs
  Tellima grandiflora (fringe-cups), p. 270

6
- Annuals with linear leaves and spikes of parchment-paper-petaled flowers
  Plantago erecta (native plantain), p. 260
- Sprawling annuals with broad, ovate, scented leaves and scattered, minute petalless flowers
  Eremocarpus setigerus (turkey or dove mullein), p. 272
- Perennial weeds from taproots with broad, ovate, unscented leaves and complex spikes or panicles of brownish to greenish flowers
  Rumex spp. (docks), p. 184

Group XI. Daisy Flowers with Both Rays and Discs
  A. Flowers yellow
  B. Flowers blue or purple (yellow centers)
  C. Flowers white (yellow centers sometimes)

Group XIA. Flowers yellow

1
- Small annuals, go to 2
- Large perennials, go to 6

2
- Flower heads bright yellow, go to 3
- Flower heads pale yellow
  Blennosperma nanum (glue-seed), p. 192

3
- Leaves and stems with sticky hairs and glands, go to 4
- Leaves and stems not glandular, go to 5

4
- Lower leaves usually lobed or divided, go to 4a
- Lower leaves entire
  Madia spp. (tarweeds), p. 198

4a

- Mostly spring flowering; floral bracts completely wrap around base of ray flowers (open flower head carefully)
  Layia spp. (tidy-tips and relatives), p. 200
- Mostly summer flowering; floral bracts wrap only part way around base of ray flowers (open flower head carefully)
  Hemizonia spp. (tarweeds), p. 198

5

- Rays inconspicuous
  Bidens frondosa (beggar-ticks)
- Rays easy to see, go to 5a

5a

- Flower heads over an inch across; clear yellow
  Monolopia spp. (no common name), p. 202
- Flower heads over an inch across; yellow with white tips
  Layia platyglossa (tidy-tips), p. 200
- Flower heads usually less than an inch across; golden yellow to yellow-orange
  Lasthenia spp. (goldfields), p. 200

6

- Flower heads no larger than a pea; densely clustered in spikes or panicles
  Solidago spp. (goldenrods), p. 202
- Flower heads larger, go to 7

7

- Plants not woody at base, leaves not wooly, go to 8
- Plants woody at base, back of leaves covered with wooly or feltlike hairs
  Eriophyllum lanatum and *confertiflorum* (wooly sunflower and golden yarrow), p. 196

8

- Heads not covered with white gum, go to 9
- Heads covered with white gum in bud
  Grindelia spp. (gumweeds), p. 196

9

- Bracts around head in one even row, go to 10
- Bracts usually in 2 or more uneven rows, go to 11

10

- Tiny ray flowers turned down
  Helenium puberulum (rosilla), p. 196
- Conspicuous ray flowers spread out
  Senecio spp. (butterwort), p. 202

11

- Leaves in large basal clumps; flowering stalks seldom more than 18 inches high, go to 11a
- Leaves along stalks several feet tall, flowers at ends, go to 11b
- Leaves along stalks to about a foot tall
  Bidens laevis (showy bur-marigold)

11a

- Leaves seldom over 3 inches long, with soft silky hairs
  Chrysopsis villosa (golden aster) p. 194
- Leaves several inches long, not silky haired, go to 11b

11b

- Floral bracts around head with stiff, ciliate hairs; pappus of 2 tapered bristles
  Helianthella castanea (Mt. Diablo sunflower), p. 198
- Floral bracts lack ciliate hairs; pappus of several scales in crown
  Wyethia spp. (mule's ear), p. 204
- Like the last but pappus is missing
  Balsamorrhiza deltoidea (balsamroot)

11c

- Disc flowers yellow, go to 11d
- Disc flowers dark purple
  Helianthus annuus (common sunflower), p. 198

11d

- Plants well over 6 feet tall, along streams or marshes
  Helianthus californicus (California sunflower), p. 198
- Plants less than 5 feet tall, on dry serpentine
  Helianthus gracilentus (serpentine sunflower), p. 198

Group XIB. Flowers blue or purple (often with yellow centers)

1

- Leaves covered with wooly hairs and ray flowers pink-purple
  Corethrogyne spp. (wooly aster)
- Plants lack this combination, go to 2

2

- Numerous linear ray flowers per head, go to 2a
- Few broader, often raggedy ray flowers per head
  Aster spp. (asters), p. 190

2a

- Leaves linear, not toothed
  Erigeron foliosus (leafy daisy), p. 194
- Leaves broader, toothed
  Erigeron philadelphicus (common daisy), p. 194

Group XIC. Flowers white (often with yellow centers)

1

- Sticky glands all over plants, go to 2
- No sticky glands, go to 3

2

- The 3 lobes of ray flowers spreading out fanlike
  Calycadenia spp. (rosinweed), p. 192
- The 3 lobes of ray flowers parallel to one another
  Hemizonia congesta luzulifolia (white tarweed), p. 198

3

- Foliage divided, often fernlike; strongly scented, go to 4
- Leaves simple, not scented, go to 5

4

- Flower heads at least ½ inch across, with yellow disc flowers in center
  Anthemis cotula (mayweed), p. 190
- Flower heads much less than ½ inch across; disc flowers whitish
  Achillea millefolium (yarrow), p. 188

5

- Leaves in basal rosettes; flower stalks only inches high
  Bellis perennis (English daisy)
- Leaves also along stems; flower stalks 1 foot or more high
  Erigeron philadelphicus (common daisy), p. 194

Group XII. Daisy Flowers with Ray Flowers Only

A. Flowers yellow
B. Flowers rose-purple or pink
C. Flowers blue or purple
D. Flowers white

Group XIIA. Flowers yellow

1

- Leaves long, grasslike on plants 3 or more feet tall
  Tragopogon dubius (salsify)
- Leaves broader, not grasslike; plants seldom over 2 feet tall, go to 2

2

- Hairs on leaves and stems with blisterlike bases
  Picris echioides (prickly ox-tongue), p. 200
- Hairs not like this, go to 3

3

- Flower heads over ½ inch across, go to 4
- Flower heads less than ½ inch across; clustered
  Lactuca spp. (wild lettuce), p. 200
- Flower heads less than ½ inch across; single at stem end
  Microseris spp. (native dandelion), p. 190

4

- Flowers bright yellow; weeds in disturbed places, go to 5
- Flowers pale yellow; weeds in disturbed places with prickly teeth on leaves
  Sonchus spp. (sow thistles), p. 200
- Flowers mostly pale yellow; natives in natural habitats, go to 6

5

- Leaves hairy; mostly 2 or more flower heads per stalk
  Hypochaeris radicata (cat's ear), p. 200
- Leaves smooth and hairless; one flower head per stalk
  Taraxacum officinale (dandelion), p. 200

6

- Flower heads nod in bud
  Microseris spp. (native dandelions), p. 190
- Flower heads upright in bud
  Agoseris spp. (mountain dandelions), p. 190

Group XIIB. Flowers rose-purple or pink

1

- Flowers red-purple, an inch or more across
  Tragopogon porrifolius (salsify)
- Flowers pale pinkish, less than an inch across
  Stephanomeria virgata (wand-lettuce)

Group XIIC. Flowers blue or purple
One species: *Cichorium intybus* (chicory), p. 194

Group XIID. Flowers white

1

- Leaves unlobed, with obvious shaggy hairs
  Hieracium albiflorum (white hawkweed), p. 198
- Leaves pinnately lobed, without shaggy hairs
  Rafinesquia californica (white chicory)

Group XIII. Daisy Flowers with Disc Flowers Only
A. Flowers yellow
B. Rose-purple or pink flowers
C. Blue or purple flowers
D. White flowers
E. Greenish or brownish flowers

Group XIIIA. Flowers yellow

1

- Plants grow on dry ground, go to 2
- Plants grow in wet, marshy spots
  Cotula coronipifolia (brass buttons), p. 194

2

- Leaves not fernlike, no special odor, go to 3
- Leaves pinnately lobed or divided, go to 2a

2a

- Plants glandular, often scented, go to 2b
- Plants neither glandular nor scented
  Chaenactis glabriuscula (yellow pincushions)

2b

- Leaves fernlike; flower heads conelike
  Matricaria matricarioides (pineapple weed), p. 202
- Leaves not fernlike; flower heads disclike, with outer disc flowers enlarged
  Lessingia germanorum (no common name)

3

- Leaves alternate, go to 3a
- Leaves opposite
  Arnica discoidea (rayless arnica), p. 190

3a

- Floral bracts spine tipped
  Centaurea solstitialis (yellow star thistle), p. 192
- Floral bracts not spiny, go to 3b

3b

- Floral bracts in one even row, go to 3c
- Floral bracts in two or more overlapping rows
  Erigeron petrophilus-inornatus complex (rayless daisies), p. 194

3c

- Flower heads one-half to one inch across; leaves simple
  Senecio aronicoides (wooly-headed butterwort), p. 202
- Flower heads much less than one-half inch across; leaves deeply pinnately lobed
  Senecio vulgaris (common groundsel)

Group XIIIB. Flowers rose-purple or pink

1

- Leaves clearly marbled with white veins
  Silybum marianum (milk thistle), p. 192
- Leaves not marbled
  Cirsium spp. (thistles), p. 192, 194

Group XIIIC. Flowers blue or purple

1

- Flower heads seldom over 1½ inches across; flowers not blue go to 2
- Flower heads very large (several inches across); flowers blue
  Cynara cardunculus (cardoon)

2

- Flower heads less than 1 inch across
  Carduus pycnocephala (Italian or plumed thistle), p. 192
- Flower heads 1 inch or more across
  Cirsium vulgare (bull thistle), p. 192

Group XIIID. Flowers white

1

- Flowers obvious (but tiny), go to 2
- Flowers not obvious; fruiting heads showy with alternating rows of white scales of two shapes
  Achyrachaena mollis (blow wives), p. 188

2

- Actual disc flowers yellow but surrounded by whitish bracts, go to 3
- Actual disc flowers white, surrounded by greenish bracts
  Baccharis douglasii (Douglas's baccharis)

- Plants making large interconnected colonies; bracts around head pearly white
    Anaphalis margaritacea (pearly everlasting), p. 196
- Plants growing as clumps; bracts around head covered with white wool
    Gnaphalium spp. (cudweeds), p. 196

Group XIIIE. Flowers greenish or brownish

1

- Leaves strongly sage scented
    Artemisia douglasiana (mugwort)
- Leaves not sage scented
    Erechtites prenanthoides (Australian fireweed)

Group XIV. Flowers with no sepals or petals.

This is such a small group, there are no separate color categories.

1

- Flowers greenish, go to 2
- Flowers whitish; forest plants
    Actaea rubra arguta (baneberry)

2

- Leaves deeply lobed to compound; plant 3 feet or more tall
    Datisca glomerata (Durango root)
- Leaves fan shaped; low sprawling plant only inches high
    Alchemilla occidentalis (western lady's mantle)

Encyclopedia of Wildflowers

ARALIA OR GINSENG FAMILY (ARALIACEAE).

Perennial herbs (shrubs and trees in the tropics), with pinnately compound leaves, sheathing at the base; flowers small and white to greenish, arranged in umbels; 5 tiny sepals, 5 petals and 5 stamens, single pistil with inferior ovary. Fruit a fleshy berry.

Elk Clover (*Aralia californica*). P. 300.

Elk clover is one of California's truly giant herbs, growing up to more than ten feet tall in a single season. Winter dormant, it produces shoots in early spring, flowers by summer, and fruits in fall. In keeping with the dimensions of the plant the leaves often reach up to several feet long. The greenish yellow flowers are borne in umbels, but despite their large number are relatively inconspicuous. The red-purple berries give a flush of color just before the plant dies back for its winter rest.

BELLFLOWER FAMILY *(CAMPANULACEAE).*

Nonwoody plants with milky juice; leaves alternate, undivided; flowers regular or irregular; 4 or 5 sepals and petals, petals joined at base, often into a short tube or bell, 5 stamens, single pistil often with 3-lobed stigma and inferior ovary. Fruit a capsule.

Annual or Rock Harebell (*Campanula exigua*). P. 301.

Rock harebell is a modest relative of the Scotch bluebell, a low annual with undistinguished leaves. One clue to identity is the milky sap that flows when a stem or leaf is broken. The open, upright pale blue bells appear in late spring and early summer, borne just inches above the rocky ground. You have to look for it during the short flowering season, for otherwise it is nearly invisible. Few other bluebells occur in our area, although *C. prenanthoides*—with nodding bells—is a common perennial species found in redwood and mixed-evergreen forests on the peninsula and in the north Bay.

Downingia (*Downingia* spp.). P. 302.

Downingias belong to the other half of the bellflower family: those with irregular, two-lipped flowers. Downingias have no satisfactorily established common name, a shame since they're among our showiest annuals. Perhaps it's because of their special habitat: vernal pools. In any event, downingias flower in the center ring of pools as pools slowly lose water from mid- to late spring. Most species have blue flowers, but the details of the blotches and spots on the middle lower petal differ for each: many have patches of white, yellow or combinations of the two, sometimes also with dark purple nipples. Each kind provides a nectar feast for native bees and each stays to its own particular vernal pool, but in areas with larger, more diverse pools two or three species may grow side by side. These are lovely wildflowers equal to garden lobelias in their beauty. It's a shame that their habitats are endangered.

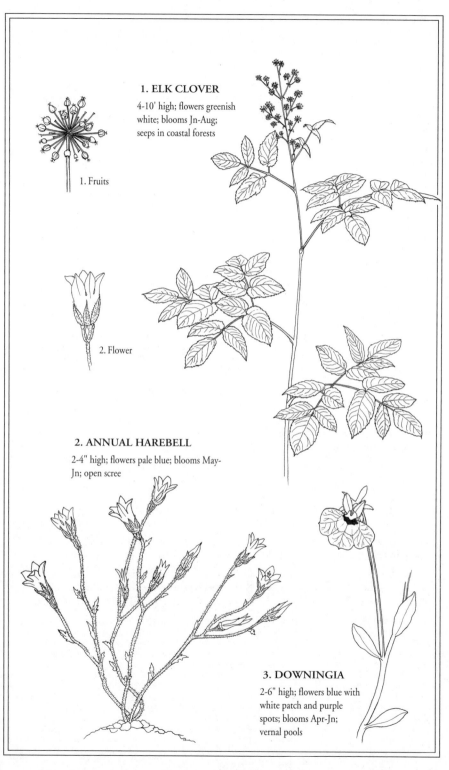

1. ELK CLOVER

4-10' high; flowers greenish white; blooms Jn-Aug; seeps in coastal forests

1. Fruits

2. Flower

2. ANNUAL HAREBELL

2-4" high; flowers pale blue; blooms May-Jn; open scree

3. DOWNINGIA

2-6" high; flowers blue with white patch and purple spots; blooms Apr-Jn; vernal pools

Looking Glass Plant (*Githopsis specularioides*). P. 303.

Looking glass plant is a seldom-seen annual growing no higher than one or two inches. It qualifies as one of our belly flowers, since you need to bend over or stretch out on your stomach to see it. Search for looking glass plant on open rocky slopes, often among short grasses and on serpentine rock outcrops. The lovely blue flowers resemble minuscule saucers, and are probably named for their resemblance to a bit of reflected sky. When pollinators are not abundant, tiny permanently closed flowers are produced close to the ground: these "cleistogamous" flowers are self-pollinating.

BIRTHWORT FAMILY (ARISTOLOCHIACEAE).

Nonwoody creeping or viny plants, often with heart-shaped leaves. Flowers borne singly, oddly shaped and colored, with 3 petallike sepals, stamens in sets of 3 and sticking to pistil; single pistil with half-inferior ovary that has 6 chambers. Fruit a capsule.

Wild Ginger (*Asarum caudatum*). P. 300.

Wild ginger carries the scent of ginger in its leaves and stems but is no relation to commercial ginger root. The latter belongs to the tropical family Zingiberaceae and comes to us from southeastern Asia, where it has been used for millenia. Our wild ginger seeks out moist forests, forming there a close ground cover as stems creep and root. The handsome, dark green, heart-shaped leaves are evergreen and completely hide flowers and fruits. The flowers are red-maroon, with long tails giving them a spidery look, and have a distinctive, sharp aroma, making them likely to be pollinated by ground-dwelling beetles. Wild ginger thrives particularly well in the moist shade of redwood forests. It also makes an excellent ground cover in woodland gardens.

BLAZING STAR FAMILY (LOASACEAE).

Nonwoody plants with undivided, rough to sandpapery leaves. Starlike yellow to orange flowers; 5 separate sepals and petals, numerous long stamens, single pistil with inferior ovary. Fruit a cup-shaped, one-chambered capsule.

Blazing Star (*Mentzelia* spp.). P. 304.

Blazing stars favor rocky slopes and disturbed places and are fond of dry areas, being especially common on our deserts. For us the showiest species is *M. laevicaulis*, which may well have been brought in by seeds in gravels used for road building and carried from their original homeland. *Mentzelia laevicaulis* (common blazing star) is a handsome biennial, with a long succession of five-pointed yellow stars a couple of inches across, opening in late afternoon and carrying through to the next morning. *M. lindleyi* (Lindley's blazing star) is another charming species uncommon here but abundant to the south. It bears deep yellow flowers on plants seldom up to two-and-a-half feet tall. Finally, *M. dispersa* is a tiny annual with minute yellow flowers to match its stature. Look for it on exposed rock scree.

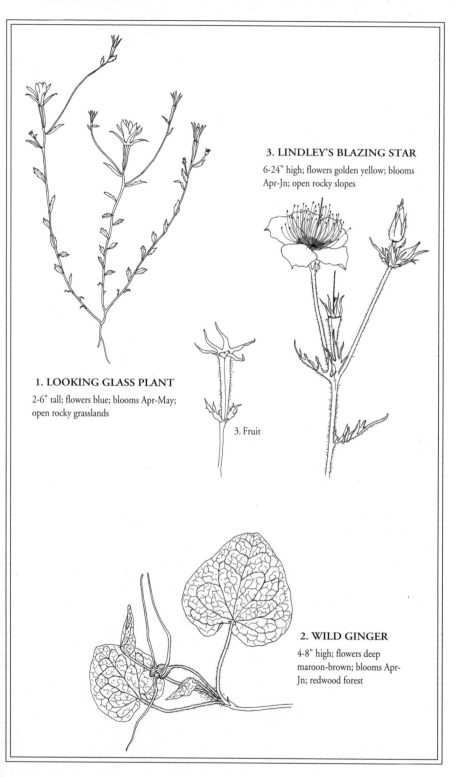

3. LINDLEY'S BLAZING STAR

6-24" high; flowers golden yellow; blooms
Apr-Jn; open rocky slopes

1. LOOKING GLASS PLANT

2-6" tall; flowers blue; blooms Apr-May;
open rocky grasslands

3. Fruit

2. WILD GINGER

4-8" high; flowers deep
maroon-brown; blooms Apr-
Jn; redwood forest

BORAGE FAMILY (BORAGINACEAE).

Nonwoody plants, often with stiff hairs on leaves and stems; leaves simple and undivided; flowers in fiddleheadlike spikes in bud; flowers of 5 sepals, 5 spreading petals fused at their bases into tubes, 5 stamens usually hidden in tube and joined to it, single pistil with superior ovary and single stigma; ovary separating into 4 lobes that ripen into nutlets in fruit.

Fiddleneck (*Amsinckia* spp.). P. 299.

Fiddleneck is named for the similarity of the young coiled-up flower spike to the fiddleneck of a violin. As the coil unwinds, flowers open; usually there is a long succession of bloom since numerous flowers occur in any fiddleneck. In good years, fiddlenecks cover the rolling foothills like a deep orange blot. Benefiting from clearing, plowing, and road cuts, fiddleneck behaves like a native weed. The various species look closely similar; details of ripe nutlets and lengths of styles are among the esoteric features used to separate them. Our most abundant species is *A. menziesii intermedia* and our rarest, *A. grandiflora*, is confined to an area near Livermore.

Popcorn Flowers (*Cryptantha* and *Plagiobothrys* spp.). P. 301, 306.

The name popcorn flower covers two genera, both of which appear alike to the novice. As with the different fiddlenecks, they're told apart by details of the nutlets revealed only with the aid of a powerful hand lens. When spring rains have come at the right time, the ground appears smothered by thousands of pieces of popcorn. Another common name is "nievitas," Spanish for "little drifts of snow." Most abundant in open fields is *Plagiobothrys nothofulvus* (pictured); many other tiny-flowered plagiobothryses bloom as vernal pools dry. Another common feature of that group is the dark purple dye produced as leaves dry. The cryptanthas are more typical of dry habitats. Their botanical name is indicative of the tiny flowers: in Greek "crypt" means hidden and "anthos" means flower. Our several species will not be detailed here.

Hound's Tongue (*Cynoglossum grande*). P. 301.

Hound's tongue is one of our earliest and loveliest woodland flowers, named for the tonguelike leaves, which unfurl through leaf duff at winter's end. At first bronze colored, they slowly turn dull green. Finally the flowering stalk emerges, and by early to mid-spring, a long succession of large, forget-me-notlike flowers open. The flowers change color: in bud they're pink-purple, open they're sky blue, a color favored by bees. The opening to the short flower tube is lined with raised white bumps (the corona), which contrast neatly with the blue petals to advertise the entrance to the tube where nectar is hidden.

BROOMRAPE FAMILY (OROBANCHACEAE).

Purple, yellow, or brownish root parasites with leaves reduced to scales. Flowers borne in dense racemes or singly, irregular, colored like the rest of the plant; petals 2-lipped, 2 above and 3 below, fused to form a tube. Other details also similar to the figwort family (Scrophulariaceae). Fruit a capsule with numerous tiny seeds.

Broomrape (*Orobanche* spp.). P. 305.

Consider yourself lucky the day you find one of these curious parasites. Looking like a fungal growth as it emerges from the soil in spring, the shoot slowly reveals flowers of color matching the scales and stem. Many species have flowers that are pretty close-up, often combining purple with yellow in a design pleasing to potential pollinators. Most orobanches are fairly specific to their host, and their roots extend a considerable distance to connect to its roots. Our few

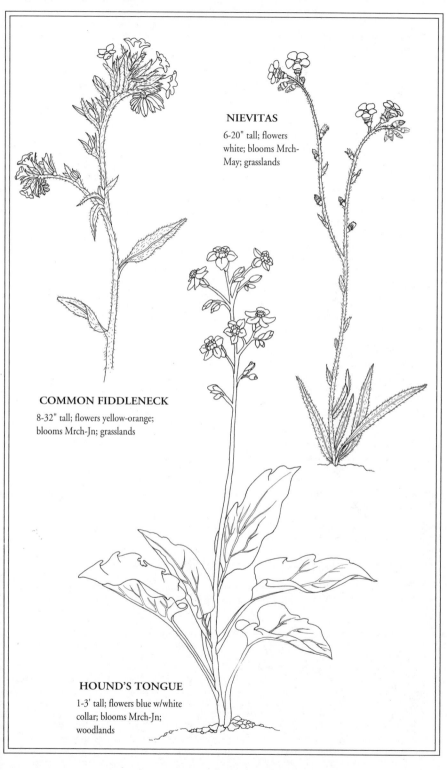

NIEVITAS

6-20" tall; flowers
white; blooms Mrch-
May; grasslands

COMMON FIDDLENECK

8-32" tall; flowers yellow-orange;
blooms Mrch-Jn; grasslands

HOUND'S TONGUE

1-3' tall; flowers blue w/white
collar; blooms Mrch-Jn;
woodlands

species vary as to color and number of flowers; the smallest is the petit *O. uniflora*, with three-inch stems bearing pretty blue-purple flowers. Others include *O. fasciculata*, with clusters of slender pale orange stems and flowers, and *O. bulbosa*, with bulbous dark purple growths covered with myriad dark purple flowers. The latter two parasitize chaparral shrubs, such as mountain mahogany and chamise. One other rare parasite in the same family is ground cone—*Boschniakia strobilacea*—whose entire flowering stalk resembles a fat, dark conifer seed cone.

BUCKWHEAT FAMILY (POLYGONACEAE).

Nonwoody or small shrubby plants with undivided leaves. Leaves sometimes bear papery stipules at their bases. Flowers small, often arranged in large spikes, panicles, or umbels, with 4 to 6 separate green or colored sepals (no petals), matching number of stamens, and single pistil with superior, usually 3-sided ovary. Fruit a one-seeded achene.

Turkish Rugging or Spine Flower (*Chorizanthe* spp.). P. 301.

Chorizanthes are seldom conspicuous unless growing en masse. Up close, these annuals are fascinating from start to finish. The start is a circular rosette of leaves, followed later by a series of upright to sprawling, forked stems that bear tiny leaves and flower clusters. Most of ours, such as the fairly common *C. membranacea*, have pale pink or rose-purple flowers, distinguished in all cases by the spiny tips on the colored sepals. Many chorizanthes continue in colorful flower even upon death, for the flowers dry nicely, retaining their original color.

Wild Buckwheats (*Eriogonum* spp.). P. 302.

Our myriad buckwheats share certain basic fundamentals. None is the edible buckwheat of commerce; that plant (*Fagopyrum esculentum*) is native to Asia; it's doubtful whether Native Americans ate seeds of our own so-called buckwheats. Leaves are usually wool covered beneath—an adaptation to reduce water loss—and are spoon shaped or rounded. Flowers are densely clustered inside vaselike or bowl-shaped floral bracts (the involucre), each flower on a threadlike stem with colored sepals only. Colors range from white through all shades of pink, rose, and red, and are also well represented in the yellows. The two buckwheats described separately are perennial, but many, especially from dry areas, are tiny annuals, such as *E. vimineum* (with dark pink flowers). Another conspicuous species—*E. wrightii*, Wright's buckwheat—appears as tight-matted mounds of white, wool-covered leaves along Mines Road. This species has all the earmarks of high montane species normally seen near and above timberline.

Naked Buckwheat (*Eriogonum nudum*). P. 302.

Naked buckwheat joins a genus that bulges with species; over eighty are known, to California's far corners. Yet of all of them this one has the widest range, extending from rocky hillsides—just a few feet above sea level and near the coast—inland across hills and mountains and climbing to timberline in the Sierra. Although variable, naked buckwheat is always typified by neat basal rosettes of spoon-shaped leaves, often wrinkled and dull green above, and white and wool-covered beneath. A single leafless stem carries flowers arranged in tight pompons on short side branches. In our area flowers are usually white, but elsewhere they may be pink or yellow. Flowers may bloom late into fall, even up until the first frosts.

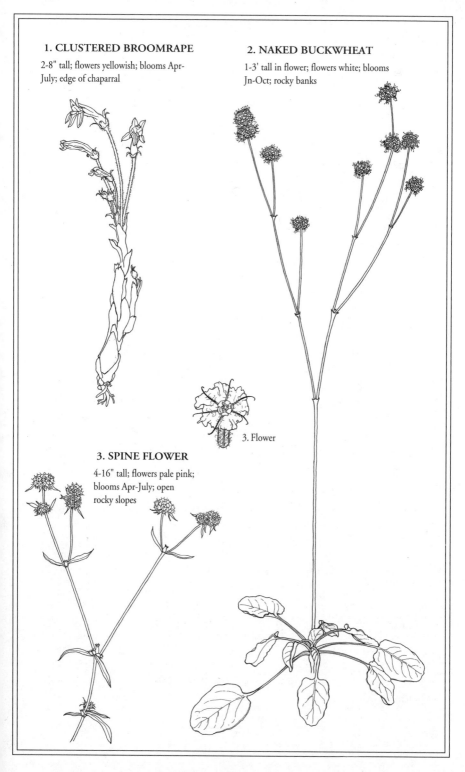

1. CLUSTERED BROOMRAPE

2-8" tall; flowers yellowish; blooms Apr-July; edge of chaparral

2. NAKED BUCKWHEAT

1-3' tall in flower; flowers white; blooms Jn-Oct; rocky banks

3. Flower

3. SPINE FLOWER

4-16" tall; flowers pale pink; blooms Apr-July; open rocky slopes

Sulfur Buckwheat (*Eriogonum umbellatum*). P. 302.

Sulfur buckwheat is at once recognizable in flower by the numerous tight mounds of sulfur yellow blossoms that enliven rocky slopes in summer. Flowers are red in bud and fade red; in fact, most buckwheat flowers dry nicely with pretty colors. Although relatively uncommon in our area—being at the lower end of its altitudinal range—sulfur buckwheat is one of the most characteristic flowers of California's mountains and high deserts, with numerous subspecies and varieties. It makes a wonderful rock garden plant for xeriscaped gardens.

Dock (*Rumex* spp.).

Docks are familiar members of that interesting urban category: "vacant lot." Most are introduced weeds from the Old World, and thrive on neglect and newly cleared land. Many features adapt docks well to this life style: long, carrotlike taproots from which they resprout when cut; broad, efficient leaves that lie close to the ground and so are hard to mow off; dense spikelike clusters of wind-pollinated greenish flowers that quickly ripen their fruits; fruits dispersed by tiny warts on the outside (resembling rice grains) that may adhere to fur or clothing (others lacking these may have wind-dispersed fruits). Perhaps the most noticeable stage in the life cycle is at fruiting time, when the three thin, winglike sepals closely invest the actual fruits and ripen to a deep coffee brown.

Sheep Sorrel (*Rumex acetosella*).

Sheep sorrel is another abundant weed from the Old World and rather distinctive in its genus. In spring, the first thing you notice are the modest clusters of arrowhead-shaped leaves, which seem to pop up here and there; actually the plants are interconnected underground by creeping runners. By April, the modest flower stalks have produced a profusion of tiny, deep-red flowers that en masse lend striking color to the landscape—usually on grassy hills or in meadows. Were it not for its wandering ways this alien might endear itself, for its sour-tasting leaves remind us of its close sister, the French sorrel (*R. acetosa*), widely used in gourmet soups. But the underground runners are almost impossible to unearth and extricate from well-established colonies, and so once planted sheep sorrel imposes year after year in gardens and vacant lots.

BUTTERCUP FAMILY (RANUNCULACEAE).

Nonwoody plants, leaves often lobed or divided into 3s; flowers variable, regular to irregular; 5 sepals (sometimes colored to replace petals), 5 petals (or missing), numerous spirally arranged stamens, several separate pistils. Fruit usually an achene or follicle.

Red Columbine (*Aquilegia formosa*). P. 300.

The slender stalks of columbine hold exquisite pendulous red and yellow flowers, which provide a visual feast for hummingbirds. The scarlet sepals are turned outwards, and the hollow yellow petals taper into knobbed, red spurs where the nectar is held. The form of the flower is said to be reminiscent of doves dipping down to drink ("columbo" is Latin for dove), while the botanical name comes from the allusion to eagles' claws. Look for red columbine along shaded streamsides growing between rocks, with saxifrages and monkeyflowers. Out of blossom, the coarsely divided leaves are attractive by themselves. Red columbine naturalizes well in the woodland garden.

2. CURLY DOCK

2-3½' tall; flowers greenish; blooms throughout year; disturbed places

2. Flower detail

1. SULFUR BUCKWHEAT

4-12" tall in flower; flowers sulfur yellow; blooms Jn-July; rocky slopes

1. Fruit

3. SHEEP SORREL

4-16" tall; flowers red-brown; blooms Mrch-Aug; disturbed places & grasslands

4. RED COLUMBINE

1½ -3' tall; flowers red and yellow; blooms May-Aug; forests

Red or Scarlet Larkspur (*Delphinium nudicaule*). P. 302.

Our second member of the buttercup family is also adapted to hummingbird pollination but plays its cards a bit differently. Here the five sepals provide most of the scarlet color: the upper sepal points backward into a single nectar-bearing spur. The smaller petals hunch inside just over the stamens. If you peel off the outer layer of the spur, however, you'll find that the two upper petals form an inner pair of spurs. This double jacket is no doubt to foil would-be nectar robbers from drilling through the spurs to reach nectar without carrying pollen. Look for red larkspur on mossy rocks in company with rock ferns, mosses, saxifrages, and stonecrop.

Larkspurs (*Delphinium* spp.). P. 302.

Our other larkspurs mostly come in shades of rich blue and blue-purple, and many look similar. Identifying features for species include whether the root is long, woody, and firmly attached to the stem or consists of a cluster of easily detached tubers. These beautiful wildflowers should never be dug, however, so it's best to appreciate them as they are. Most have deep blue sepals and less showy petals (including *D. patens*, *D. hesperium*, and the largest-flowered of all, royal larkspur, *D. variegatum*). One other species is striking for its unusual color: *D. californicum* (pictured) has pale, near-whitish flowers flushed greenish or pale lavender and is conspicuously fuzzy haired. Most larkspurs have shorter spurs than the red kind, and are pollinated by bumblebees and butterflies.

Rue-anemone (*Isopyrum* spp.). P. 303.

These are special woodland flowers that are seldom noticed, for they're delicate and have a short floral life. Out of flower the low, three-part leaves are easily overlooked. Never truly common, rue-anemones favor mossy rocks in oak woodlands, where their delicate whitish flowers open in April. If you've ever seen the rue-anemones from eastern hardwood forests, these isopyrums will look familiar. The reason they're not classified as true anemones is because the fruit is a slender many-seeded follicle and not a one-seeded achene.

Water Buttercup (*Ranunculus aquatilis*). P. 306.

When people first see the water buttercup, they have a hard time believing that it belongs to the same genus as other buttercups: the flowers are snow white rather than shiny yellow, and the rest of the plant is submerged in a shallow pool or sluggish stream. If you peer into the water you see its unique adaptation to aquatic life; leaves are dissected into long, narrow strands for superior light absorption and increased surface area to absorb carbon dioxide from the water (most plants get this gas from the air). So dependent are the leaves on their aquatic environment that they wither and dry up if exposed to the air for long. Water buttercup transforms the water like so many snowflakes as the flowers open.

Buttercups (*Ranunculus* spp.). P. 306.

We have several species of ordinary buttercups, among the ranks of the prettiest early spring flowers. Buttercups are typified by yellow petals with a varnished appearance, numerous yellow stamens, and numerous tiny green pistils in the center. The leaves are lobed like crows' feet, accounting for another common name, "crowsfoot." The genus name alludes to another trait: "Rana" is frog and "-unculus" means little, because buttercups may live in damp places where little frogs are found. Our most common species is *R. californicus* (California buttercup), abundant on grassy slopes or in open woodlands. It's our only species with ten or more petals (others have five). We also have nonnative buttercups, generally with prickles on the seed pods (hence their easy dispersal to new areas).

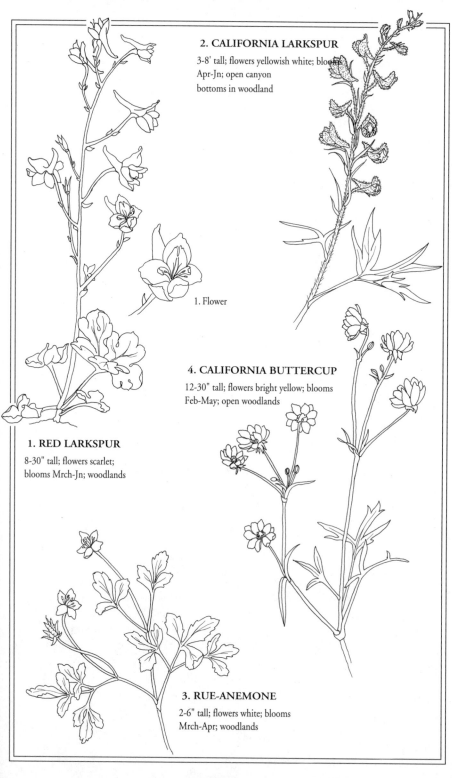

2. CALIFORNIA LARKSPUR

3-8' tall; flowers yellowish white; blooms
Apr-Jn; open canyon
bottoms in woodland

1. Flower

4. CALIFORNIA BUTTERCUP

12-30" tall; flowers bright yellow; blooms
Feb-May; open woodlands

1. RED LARKSPUR

8-30" tall; flowers scarlet;
blooms Mrch-Jn; woodlands

3. RUE-ANEMONE

2-6" tall; flowers white; blooms
Mrch-Apr; woodlands

Foothill Meadow Rue (*Thalictrum fendleri* var. *polycarpum*). **P. 307.**

The prettily divided, fernlike leaves of meadow rues are common sights in high mountain meadows, but we also have a foothill species gracing rocky streamsides. Dormant through fall and winter, the new leaves unfurl in spring, followed by the somewhat disappointing flower stalks. Flowers have green sepals only—no petals—and male flowers are borne on plants separate from the female, assuring cross-pollination by winds. At close range, the male flowers look like little green umbrellas with dozens of dangling stamens; the female, green saucers with four or five separate pistils that later ripen into achenes.

DAISY OR COMPOSITE FAMILY (ASTERACEAE).

Nonwoody plants or shrubs with variable leaves. Many small flowers arranged in dense heads to resemble single, larger flowers. Head surrounded by sepallike floral bracts; flowers inside may be of one or two types; ray flowers are showy, with single strap-shaped petals, and are often sterile; disc flowers are small, with tubular petals ending in a 5-lobed starlike design. Sepals are modified into a pappus consisting of hairs, bristles, plumes, or scale-shaped appendages that help disperse the one-seeded fruits; 5 stamens, fused together by their anthers; one pistil with inferior ovary and 2-branched style. Fruit a one-seeded achene. This large family is often subdivided into tribes. A good hand lens is especially helpful in looking at flower parts; you'll have to pull open the flower head to find the pappus, or wait until it reaches the fruiting stage.

Yarrow (*Achillea millefolium*). **P. 299.**

Yarrow is among the most widespread of the aromatic daisy relatives, spanning the northern hemisphere. Even in California, yarrow ranges from seashore to mountaintops, where it seeks open grassy or rocky habitats. The finely divided, fernlike leaves give rise to the species name "millefolium" meaning thousand leaves. These leaves are endowed with strong, bitter oils—with the aroma of sage—and have been used medicinally since time immemorial. The flat-topped clusters of flowers are really clusters of flower clusters, since each smallish white "flower" is really a miniature head of flowers, each complete with a few white ray flowers and several creamy disc flowers (use a good hand lens). The strong odor is an adaptation to discourage insect marauders and would-be browsers. Yarrows are also being used in natural wild garden designs or as low-maintenance ground covers as a droughty substitute for grass.

Blow-wives (*Achyrachaena mollis*). **P. 299.**

Blow-wives is one of those strange, unforgettable names, which may not have much meaning. This spring annual is a tarweed relative that is seldom noticed in flower, for the long floral bracts extend snuggly up around the blossoms. Instead the pale yellow flowers are self-pollinated, a trait often seen in weedy plants, although blow-wives is seldom a garden pest. The attractiveness of this plant lies instead with the one-seeded fruits, each of which wears two rows of white, satiny, crownlike scales, a longer one alternating with a shorter one, giving the impression that these are double-petaled flowers. The floral bracts have meanwhile turned back so that these pretty fruits are fully revealed, but not for our benefit; rather this allows the wind to pick up fruits by their white sails and "blow" them away.

1. Female flowers

1. Male flowers

1. FOOTHILL MEADOW RUE

2-6' tall in flower; flowers green; blooms Mrch-May; canyons in woodlands

3. BLOW-WIVES

4-16" tall; flowers yellow; blooms Apr-May; open places & grasslands

2. YARROW

1-3' tall in flower; flowers white; blooms Mrch-Jn; open places

Native or Mountain Dandelions (*Agoseris* and *Microseris* spp.). P. 299, 304.

The beginner is often confused by the plethora of dandelionlike flowers found in the wild. Some are true weeds and alien to our area; others are native wildflowers that seldom act weedy. The agoserises and microserises belong to the latter category. They favor grasslands, open woodlands, sandy soils, or rocky slopes in half-shade according to kind. The leaves vary from narrow and grasslike to lance shaped, but they are cut deeply or have prettily recurved lobes. It's not really necessary to know all the kinds, but generally the genus *Agoseris* has flower heads that are held upright from bud through seed, while the genus *Microseris* has nodding buds and upright flowers. Both may call attention to themselves a second time when the plumes of white hairs or silvery scales covering the fruits open out to catch the wind, much in the manner of their unwanted weedy cousins.

Mayweed or Dog Fennel (*Anthemis cotula*).

Mayweed announces its relationship to the herb called chamomile by the deeply slashed, fernlike foliage and the pretty small white and yellow flower heads: white rays around a cone-shaped cluster of minute yellow discs. One whiff of the crushed leaf, however, and you're likely to suspect that this is not a suitable plant from which to brew tea; the smell is generally unpleasant. Nonetheless, this would-be wildflower is an interloper from Europe—the same homeland as for chamomile—and makes itself apparent in late spring when temporarily wet, grassy fields are covered with myriad flowers.

Rayless or Coastal Arnica (*Arnica discoidea*). P. 300.

In the Old World, arnicas were known for their reputed medicinal properties. Arnicas in California are associated with memories of mountain strolls, for most species are found on the edge of mountain meadows or under pine woods. Yet rayless arnica is likely to go unnoticed, for although it's scattered through the Coast Ranges, its flower heads lack the bright golden rays of most mountain sisters. Instead each head—nodding in bud—consists of a single row of hair-covered floral bracts surrounding a modest head of yellow disc flowers. One look at the leaves, however, and you know it's an arnica, for the leaves are broadly oval shaped and paired, as are those of so many mountainous species.

Asters (*Aster chilensis* and *radulinus*). P. 300.

"Aster" is a version of the word "star," since the purple, blue, or white rays look like points of a yellow-centered star. Although well-known in the flora of the eastern United States for their profusion of bloom at summer's end, our native asters often go unheeded. Most occur in high mountains, and many asterlike flowers belong to a related genus, *Erigeron* (known as fleabane or daisy). Nonetheless we have two common asters: *A. chilensis* is a vigorous perennial with creeping, spreading roots, narrow leaves, and open clusters of flower heads with pale bluish rays. Look for it in fall in temporarily wet ditches or swales. *A. radulinus* is called broadleaf or woodland aster, and lives in open oak and bay woods, where its broad, toothed leaves remain a mystery out of flower. At summer's end it sends up a few white- to pale-purple-rayed flowers.

1. MOUNTAIN DANDELION

5-24" tall in flower; flowers yellow; blooms
May-Jn; open woodlands

2. MAYWEED

4-20" tall; flowers white
w/yellow center; blooms
May-July; disturbed fields

3. Bracts

4. BROADLEAF ASTER

8-24" tall; flowers white w/ pale
yellow center; blooms July-Oct;
woodlands & dry forest

3. RAYLESS ARNICA

1-2' tall; flowers yellow;
blooms May-July; woodlands

Glue-Seed (*Blennosperma nanum*). P. 300.

Many of our small bright yellow daisies at first glance look alike, but glue-seed stands apart by the pretty pale yellow color of its daisies, further accented by the blobs of white pollen coming out of the stamens (a good, strong hand lens is useful here). In addition, glue-seed blossoms before many of the other daisies, from early to mid-spring, while local depressions in grasslands still are muddy, or it occurs along the edge of vernal pools. These pretty flowers have earned the name glue-seed for the fact that their tiny fruits glue themselves to clothing or fur when wetted, as so often happens during spring rains, and this provides an efficient means of dispersal.

Rosinweed (*Calycadenia* spp.). P. 300.

Rosinweed is a member of the tarweed group and, like most tarweeds, excels at flowering and seeding when conditions are at their worst—bone dry soils in summer—even up to summer's end. Add to this the difficulty of surviving on nutrient-poor serpentine soils, and you've got a truly remarkable wildflower. The rosinous covering of glands on stems and leaves helps seal water in and keep munching critters out, but this does not fully explain its remarkable ability to bloom when all is sere. Rosinweeds are immediately identified by their white to pale pink ray flowers. Their three lobes fan out in different directions, giving the flower heads their own peculiar and distinctive appearance.

Nonnative Thistles (various genera and species).

Since most thistles look similar because of their spine-tipped leaves, stems, and flower heads, all the weedy, nonnative kinds are discussed together under one heading. Thistles have found heaven in California's roadsides, overgrazed pastures, and disturbed fields, since they have carte blanche to grow, flower, and seed without being munched or disturbed. Their spines are formidable weapons against being eaten, and even the gardener must be wary when uprooting these plants. In addition to their spininess, all thistles have compact heads of extra long, tubular disc flowers. These are favored by bumblebees and butterflies, which are their principal pollinators. Our area is home to a wide variety of nonnative thistles, the boldest of which is the artichoke relative called cardoon (*Cynara cardunculus*), with enormous heads of blue-purple flowers. Perhaps the Italians first brought in cardoon: the leaves are often eaten as a vegetable. Next in size are two stout, tall thistles with flower heads only a couple of inches across: bull thistle (*Cirsium vulgare*), with rose-purple flowers and solid green leaves; and milk thistle (*Silybum marianum*), with similar purple flower heads but leaves beautifully veined and marbled with white. A smaller-headed but still rank and tall thistle is the Italian or plume thistle (*Carduus pycnocephala*), with one-half-inch heads of pale purple flowers. Italian thistle is particularly well armed with nasty spines. Finally, the short but still pernicious yellow star thistle (*Centaurea solstitialis*) has unspiny leaves and stems, but the bright yellow flower heads —favored by bees—are surrounded by long, radiating spines. Star thistle has become a major problem to grazing livestock, containing a poison that sickens and eventually kills them. Research is being done to introduce one or more natural pests from star thistle's Mediterranean homeland to control its rampant spread.

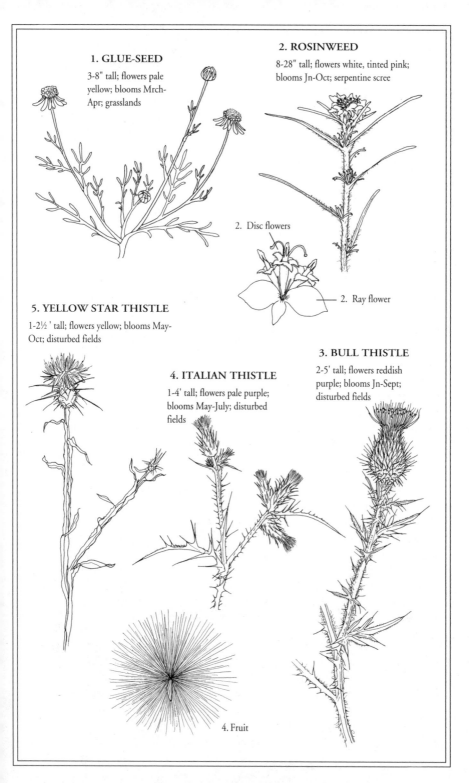

1. GLUE-SEED

3-8" tall; flowers pale yellow; blooms Mrch-Apr; grasslands

2. ROSINWEED

8-28" tall; flowers white, tinted pink; blooms Jn-Oct; serpentine scree

2. Disc flowers

2. Ray flower

5. YELLOW STAR THISTLE

1-2½ ' tall; flowers yellow; blooms May-Oct; disturbed fields

4. ITALIAN THISTLE

1-4' tall; flowers pale purple; blooms May-July; disturbed fields

3. BULL THISTLE

2-5' tall; flowers reddish purple; blooms Jn-Sept; disturbed fields

4. Fruit

Native Thistles (*Cirsium occidentale* and varieties). P. 301.

Although close relatives of bull thistle and other nonnative thistles, the native thistles behave in different fashion, growing in small groups or singly on rocky hillsides away from disturbance. One look at them, however, and there's no question they're thistles through and through, for they also have spine-edged leaves and floral bracts that end in sharp thorns. Despite this, there is more than subtle beauty as the flower heads open to reveal rose-purple, pink, or crimson-red blossoms that lure both hummingbirds and butterflies with their brilliant colors and copious nectar. These thistles are often referred to as cobweb thistles owing to the cobwebby hairs that are festooned over floral bracts and the undersides of leaves.

Chicory or Blue Sailors (*Cichorium intybus*).

No other roadside wildflower elicits such compliments, for its morning flowers are elegant pale sky-blue dandelions arrayed on tall bushy plants. Yet out of flower chicory's numerous scrawny, armlike branches with ratty leaves are hardly pretty. Chicory is not native, no matter how widely distributed along California's summer-dry roadsides, but comes from the Mediterranean, where its leaves have long been used for bitter greens and its roots roasted as a coffee additive. The endive of commerce is a specially selected kind of chicory that is grown in the dark so its new leaves remain tender and sweet.

Golden Aster (*Chrysopsis* [*Heterotheca*] *villosa*). P. 301.

The true asters described above show a contrast between yellow discs in the center and purple, blue, or white rays, a color combination irresistible to bees. Golden aster differs by appearing all yellow (at least to the human eye), but doubtless there is an ultraviolet design superimposed over the yellow, since bees see ultraviolet coloring well. In fact, bees see blues, purples, and yellows but not the red end of the spectrum. Golden aster is a low, unassuming, semimatted rock plant unnoticed until it's covered with myriad golden daisies in early summer.

Brass Buttons (*Cotula coronipifolia*).

Some plants specialize in occupying very specific niches; brass buttons is one of these, made more surprising since it's native to South Africa. Look for it in muddy places or in standing, brackish water, where it has little competition. The fleshy leaves remind us that extra water is often stored in plants where salt is in the soil. The small flower heads are shaped like buttons but are a clear bright yellow and consist only of tiny disc flowers.

Common Daisy (*Erigeron philadelphicus*). P. 302.

Despite its "common" name, the common daisy is not particularly abundant in our area; its name comes rather from the fact that it is widespread across the United States. Look for common daisy in temporarily moist places in grasslands. There it produces a long succession of typical daisies, with numerous narrow white, pale purple, or pink ray flowers with yellow centers. Daisies are easily confused with asters. Look closely at their floral bracts; in daisies their tips are generally black, brown or purplish, while asters' bracts are green tipped. Another difference is the numerous, narrow ray flowers of daisies as contrasted with the few, ragged ray flowers in asters. In addition to common daisy, we have a few other erigerons: *E. foliosus* (leafy daisy) has slender stems clothed with nearly linear leaves and modest flower heads surrounded by blue-purple rays; *E. petrophilus* (rock daisy) and *E. inornatus* (rayless daisy) are a closely related pair with similarly leafy stems but flower heads lacking rays altogether.

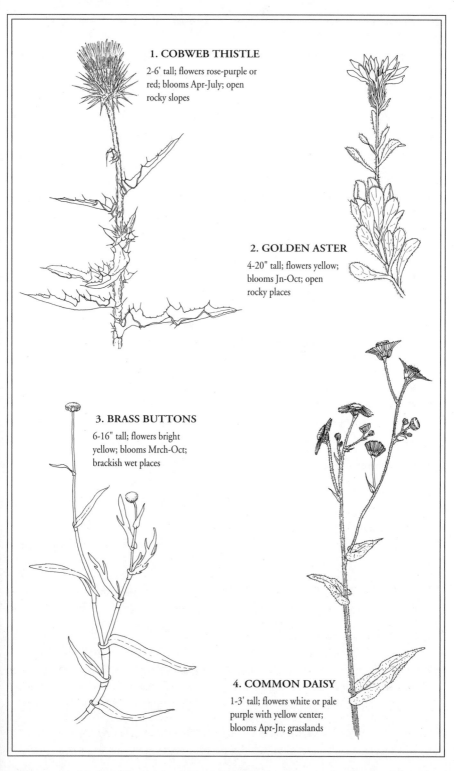

1. COBWEB THISTLE

2-6' tall; flowers rose-purple or
red; blooms Apr-July; open
rocky slopes

2. GOLDEN ASTER

4-20" tall; flowers yellow;
blooms Jn-Oct; open
rocky places

3. BRASS BUTTONS

6-16" tall; flowers bright
yellow; blooms Mrch-Oct;
brackish wet places

4. COMMON DAISY

1-3' tall; flowers white or pale
purple with yellow center;
blooms Apr-Jn; grasslands

Golden Yarrow and Wooly Sunflower (*Eriophyllum confertiflorum* and *lanatum*). P. 302.

The eriophyllums are named from two Greek words: "erios" for wooly, referring to the densely wooly hairs on stems and undersides of leaves; and "phyllos" for leaf. Wooly hairs on the undersides of leaves prevent them from drying out. This allows eriophyllums to grow in decidedly dry habitats, usually rocky hillsides or openings in chaparral. The two species use different strategies for attracting pollinators to their golden-rayed and -centered flowers: golden yarrow masses many small flower heads together into flat-topped clusters for a dramatic visual effect, whereas wooly sunflower sends up single, large flower heads one at a time. It's difficult to say which is more successful; actually, the two species seldom grow together.

Cudweeds and Pearly Everlasting (*Gnaphalium* spp. and *Anaphalis margaritacea*). P. 303, 299.

Both the cudweeds and pearly everlasting are members of the everlasting tribe, in which it's common for flower heads to dry and retain their color. The two are also similar in making a show of the dry, pearly white to pale brown or pinkish floral bracts that substitute for the lost ray flowers. Inside these bracts the tiny yellow disc flowers are scarcely noticed without the aid of a hand lens, yet small insects must discover and pollinate them, for the widely dispersed seeds are clearly successful. Cudweeds and pearly everlasting also have similar leaves and stems, covered with cobwebby to wooly hairs, an adaptation to prevent drying out. To tell the two apart, observe how they grow. Pearly everlasting creates large interconnected colonies through creeping roots; cudweeds spring from single taproots. The several kinds of cudweed are too detailed for this book; suffice it to say that some are aliens from South America and Europe while others are well-behaved natives. Many have alluring smells when their leaves are crushed; a couple of species have overtones of curry. Illustrated is a handsome native species, *G. californicum*, with two-foot stems and uncommonly bright green leaves.

Gumweed (*Grindelia* spp.). P. 303.

The name gumweed makes this attractive daisy sound undesirable, but in fact it is one of the prettiest yellow daisies of summer: flower heads may measure over an inch across, with myriad bright golden ray flowers and matching central discs. Well-endowed with nectar, they are delights to bees and many other pollinators. The common name refers to the white, gummy substance that covers the flower buds, a means of discouraging would-be chewers from getting at the flowers before they open. This same trait makes identification of the genus easy, but the several species are difficult to distinguish. Gumweeds are also striking for their ability to grow under adverse circumstances: near the coast they may creep over sand dunes or occupy the upper end of salt marshes; in our area they're more likely to grow on stony soils in open fields or along roadsides.

Rosilla (*Helenium puberulum*). P. 303.

Rosilla is a unique but unprepossessing plant; you need to know where and what to look for. Favoring permanent seeps or the edges of small streams, the plants have tall stems that may bear flowers all through the summer, though these hardly call attention to themselves. They consist of a dome-shaped cluster of dark brownish disc flowers surrounded at the base by tiny yellow, turned-down rays. The unusual leaves are the true signal of this plant, for the blade continues down the stem as a wavy green wing below the node. Perhaps the common name—from Spanish for light red—alludes to the fact that these flowers are the source of a reddish dye.

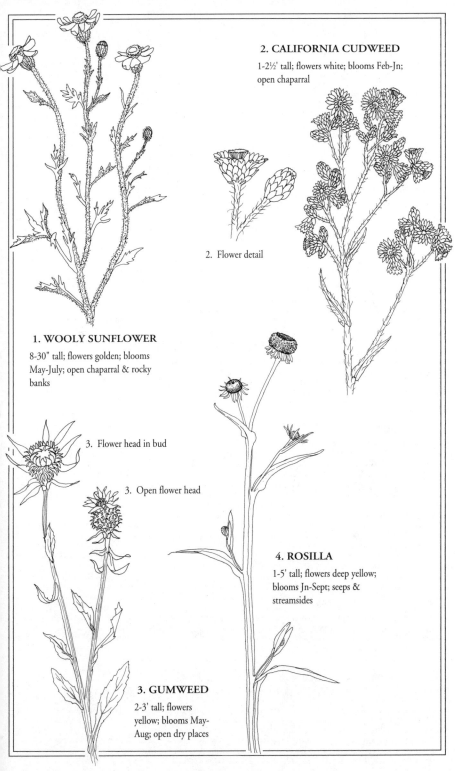

2. CALIFORNIA CUDWEED

1-2½' tall; flowers white; blooms Feb-Jn; open chaparral

2. Flower detail

1. WOOLY SUNFLOWER

8-30" tall; flowers golden; blooms May-July; open chaparral & rocky banks

3. Flower head in bud

3. Open flower head

4. ROSILLA

1-5' tall; flowers deep yellow; blooms Jn-Sept; seeps & streamsides

3. GUMWEED

2-3' tall; flowers yellow; blooms May-Aug; open dry places

Mt. Diablo Sunflower (*Helianthella castanea*). P. 303.

Many rare wildflowers are not particularly distinguished for their looks, but the Mt. Diablo sunflower is a clear exception. Restricted to the region around the mountain, this sunflower forms low, dense cushions of handsome dark green leaves, then carries showy yellow flower heads singly on stalks well above the leaves. Look for these cheerful harbingers of summer in chaparral clearings or by rocky outcrops from the middle elvations of the mountain to the top. A closely related and widespread species—*H. californica*—is neither so conspicuous nor so large flowered and apparently does not enter our area.

Sunflowers (*Helianthus annuus* and *californicus*). P. 303.

Sunflowers are as familiar as any native wildflower, partly because the midwestern kinds are familiar in a giant form grown for its oily, edible seeds. It is called sunflower because the stem turns to follow the sun. Our area has three very different species: The annual nonnative sunflower (*H. annuus*) is a common sight along roadsides and farm fences through the summer, with showy heads of yellow rays and dark black-purple disc flowers. It comes to us from the Midwest. The indigenous California sunflower (*H. californicus*) is a giant perennial plant growing on banks of temporary streams. It has smaller flower heads borne in profusion, with rays and discs both yellow, and it has starchy rhizomes that increase their tenure on the land each year. In this it resembles another sunflower—*H. tuberosus*—better known to gardeners as Jerusalem artichoke. Finally the serpentine sunflower (*H. gracilentus*) is a less robust native found on dry, rocky soils where there's little competition with other plants.

Tarweeds (*Hemizonia* and *Madia* spp.). P. 303, 304.

The tarweeds have an undeserved lowly reputation. In truth, most are native and able to grow under some of the most difficult conditions known—bone dry, cement-hard adobe soils in hot summer sun. Despite this, many have attractive and colorful flowers. The bad rap is on two counts: the (to some) disagreeable odor and tarry glandular substances that cover their stems and leaves (and incidentally allow them to survive severe drought); and the fact that many grow with great abandon in pastures and along roadsides. If instead you think of them as valiant survivors, actually beginning active growth only in summer—when almost everything else is dying or going dormant—and flowering and seeding as late as fall, you'll see what marvels they really are. Of the several species and two major genera in our area, hemizonias tend to have their lower leaves deeply slashed or divided, while madias have simple, undivided foliage. Some species bear small yellow daisies, others larger pure white daisies (such as *Hemizonia congesta luzulifolia*, a spring bloomer). Still others have uncommonly large yellow daisies, centered deep crimson. The latter description belongs to *Madia elegans*, the elegant tarweed, which opens its colorful blossoms only in late afternoon, evening, and early morning unless fogs keep the skies gray. In full flower, it is the equal of any garden daisy. *Hemizonia fitchii* is another example of a local tarweed, this one with rigidly spine-tipped leaves for extra protection. The bright yellow flower heads on this kind are pretty but smaller than for elegant madia and are wide open in the middle of the day.

White Hawkweed (*Hieracium albiflorum*). P. 303.

Unassuming any time of the year, white hawkweed is one of our most common forest and woodland wildflowers. Out of flower, it has rosettes of long, elliptical leaves with conspicuous, shaggy white hairs; later, the flowering stalk carries several miniature pure white dandelionlike heads, consisting of ray flowers only. Even though white hawkweed has "weed" in its name, it is a native and seldom behaves in weedy fashion, staying put in its native haunts.

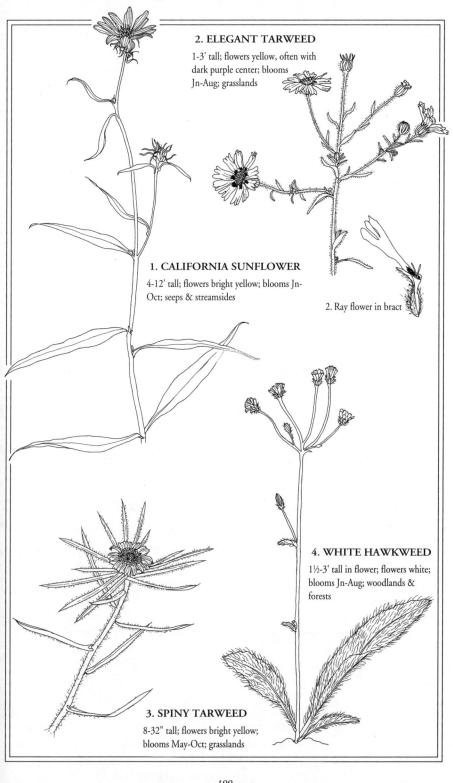

2. ELEGANT TARWEED

1-3' tall; flowers yellow, often with
dark purple center; blooms
Jn-Aug; grasslands

1. CALIFORNIA SUNFLOWER

4-12' tall; flowers bright yellow; blooms Jn-
Oct; seeps & streamsides

2. Ray flower in bract

4. WHITE HAWKWEED

1½-3' tall in flower; flowers white;
blooms Jn-Aug; woodlands &
forests

3. SPINY TARWEED

8-32" tall; flowers bright yellow;
blooms May-Oct; grasslands

Nonnative Dandelions and Relatives (*Hypochaeris* spp., *Taraxacum officinale*, and *Picris echioides*).

The several nonnative dandelions are closely related but are placed in different genera; all of them come to us from the Mideast and Europe, and many are now worldwide in distribution through the activities of humans and wind. All bear heads of bright yellow ray flowers, and all have a milky juice when cut. All act in weedy fashion to invade lawns, gardens, vacant lots, and grazed grasslands. The true dandelion (*Taraxacum officinale*) is least common but crops up in lawns, where its low-lying rosettes of nonhairy, deeply notched leaves avoid being mowed because they're so close to the ground. If they are cut, the taproot quickly produces a new set of leaves. The flower heads are carried singly at the ends of short, leafless stalks. Cat's ear (*Hypochaeris radicata*) is much more abundant as a garden weed. It has a rosette of rather coarse, fuzzy leaves (hence the common name) and taller flowering stalks that bear two or more flower heads. Sow thistles (*Sonchus* spp.) are exuberant garden weeds with coarsely slashed, spine-edged leaves and small heads of pale yellow dandelionlike flowers. Prickly ox-tongue (*Picris echioides*) is the coarsest of the lot, with multiple branches bearing many smaller flower heads. It's easily recognized by the leaves covered with numerous sharp white hairs whose bases are puckered each like a raised boil (again the reason for the common name). All species produce single-seeded fruits, which in dandelion and cat's ear are crowned by white hairs, parasol fashion. Hairs facilitate dispersal to new territory, wafted by the slightest breeze.

Wild Lettuce (*Lactuca serriola* and *saligna*).

The word "lettuce" originates from the Latin word for milk (English cognate, "lactate") because the original wild lettuces—from the Mediterranean region—had copious milky juice in leaves and stems. This milky juice renders the plants intensely bitter and prevents them from being eaten. Garden lettuce has been bred so this juice is not produced until late in the life cycle; if you let your lettuce plants flower, you'll find that their leaves, which have turned bitter, also have a milky juice. Wild lettuces are bitter from the beginning—and so not appropriate for eating—but they have thoroughly established themselves as summer weeds in dry, grassy, disturbed or grazed areas. Minute yellow flower heads resemble miniature dandelions, while the leaves differ from the tight heads found on garden lettuce: *L. saligna* bears vertically oriented blue-green spoon-shaped leaves; *L. serriola* (pictured) has deeply sculpted pale green leaves.

Goldfields (*Lasthenia* spp.). P. 303.

No more appropriate common name could be given to this wildflower, which is our most abundant mid-spring yellow daisy, for these annuals cover grassy fields and rolling hills by the thousands in years with good rains. What they lack in short stature—growing only inches tall—they make up for in numbers of plants. Each flower head is a perfect golden daisy: one row of slightly curved yellow ray flowers and a large center filled with dozens of matching disc flowers. With goldfields, the bees are assured a reliable source of nectar for a couple of weeks each spring. Of our species, *L. californica* is by far the most abundant. *L. glabrata* is less common, although found on grasslands near the coast with flower heads between one and two inches across.

Tidy-tips (*Layia platyglossa*). P. 303.

Just about the time the gold is fading from the goldfields, tidy-tips begins to open its large, colorful daisies, often in the same field. Although belonging to the tarweed tribe, tidy-tips is not nearly so glandular or smelly as most tarweeds, and it blooms in mid- to late spring instead of

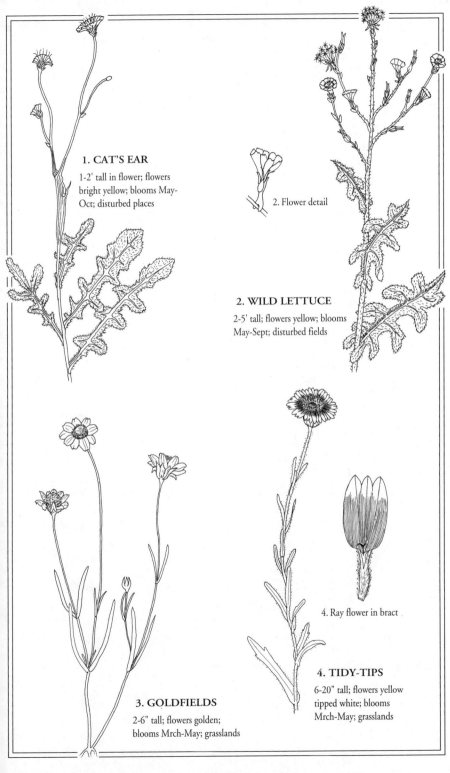

1. CAT'S EAR

1-2' tall in flower; flowers
bright yellow; blooms May-
Oct; disturbed places

2. Flower detail

2. WILD LETTUCE

2-5' tall; flowers yellow; blooms
May-Sept; disturbed fields

4. Ray flower in bract

4. TIDY-TIPS

6-20" tall; flowers yellow
tipped white; blooms
Mrch-May; grasslands

3. GOLDFIELDS

2-6" tall; flowers golden;
blooms Mrch-May; grasslands

summer. The flowers of this particular species illustrate the reason for the common name, for the broad ray flowers are golden yellow for two-thirds their length, but they are tipped white, with a clean line separating the white from the yellow. The disc flowers are the usual bright yellow, matching the base of the rays and producing in overall effect a particularly pretty color combination. From afar the eye reads the overall color combination as pale yellow, since the white and golden yellow combine.

Pineapple Weed (*Matricaria matricarioides* [*Chamomilla suaveolens*]).

Pineapple weed is a close relative to mayweed—commented on above—but without the offensive odor and without white ray flowers. Instead its finely divided, ferny leaves carry a pleasant fruity aroma, and the thimble-shaped flower heads have masses of tiny yellow disc flowers only. Related to chamomile, the dried flower heads can be readily substituted and make a tasty herbal tea. Look for pineapple weed in the worst possible habitats: hard-packed, trampled soils in vacant lots or on the borders of gardens. It wins this habitat by default.

Monolopia (*Monolopia major*). P. 305.

It's a shame this colorful yellow daisy has no well-accepted common name, for it's really one of the largest-flowered and most prolific mid-spring wildflowers, in years of generous rains. Then, normally dry banks on steep slopes or shallow sandy soils are smothered by thousands of plants in simultaneous bloom. The three-inch flower heads have showy yellow rays and centers of matching yellow disc flowers.

Senecios or Butterworts (*Senecio* spp.). P. 307.

The genus *Senecio* is one of the world's largest, being distributed all around the globe and in a wide variety of habitats. Some are adapted to deserts, others to rain or cloud forests, still others to the alpine zone of high mountains. All share a single even row of floral bracts around the head, alternate leaves, and a pappus of white hairs, which aid wind dispersal of the seeds. In fact, the word "senecio" comes from the same root as "senescent," which means old and in this case alludes to the whiteness of those pappus hairs. There are two native senecios. *S. aronicoides* is an unprepossessing short-lived perennial with tongue-shaped, unlobed leaves and spidery hairs around the flower buds. Open, the flower head disappoints, for it consists only of yellow disc flowers. *Senecio breweri*, on the other hand, has deeply pinnately lobed leaves and flower heads bordered by narrow yellow rays. At the peak of bloom, these flowers transform the plant for a brief time.

California Goldenrod (*Solidago californica*). P. 307.

Like the asters, the eastern goldenrods are well known for their late summer and fall flowers, yet California has several colorful goldenrods of its own. California goldenrod is a pretty perennial with running, spreading roots that over time create large colonies, on open banks. As with other goldenrods, the show is in the sheer numbers of flower heads: each head is about the size of a pea but complete with yellow rays and discs. Any "spike" of flowers actually carries dozens of individual flower heads, open or in bud at the same time. Flowering may last up to the first autumn frosts.

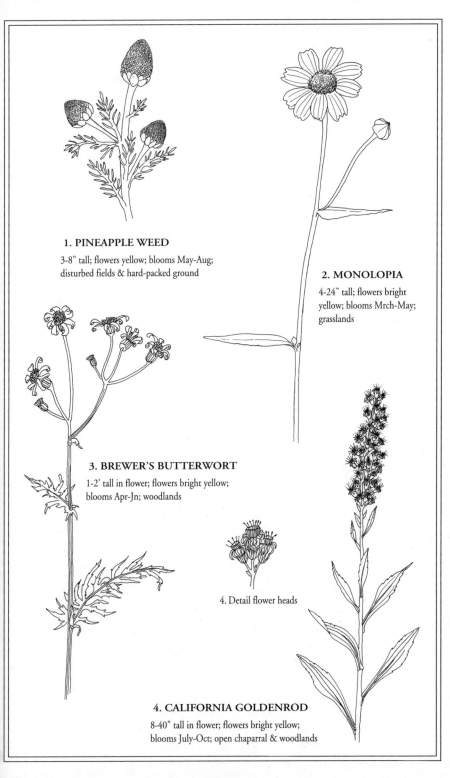

1. PINEAPPLE WEED

3-8" tall; flowers yellow; blooms May-Aug;
disturbed fields & hard-packed ground

2. MONOLOPIA

4-24" tall; flowers bright
yellow; blooms Mrch-May;
grasslands

3. BREWER'S BUTTERWORT

1-2' tall in flower; flowers bright yellow;
blooms Apr-Jn; woodlands

4. Detail flower heads

4. CALIFORNIA GOLDENROD

8-40" tall in flower; flowers bright yellow;
blooms July-Oct; open chaparral & woodlands

Mule's Ear (*Wyethia* spp.). P. 308.

Mule's ear is closely related to the sunflowers, a fact readily seen when both are in flower; in fact, mule's ear flower heads measure several inches across, vying with those of commercial sunflowers for size, and are characterized by many bright golden yellow rays and discs. The way they grow, however, is quite different. Where sunflowers have tall, straight stems, mule's ear grows as large leafy colonies close to the ground, spreading by underground stems, with basal leaves; flower heads are carried singly, slightly above these leaves. The common mule's ear (*W. glabra*) is named for its broad, floppy, earlike leaves, while the narrow-leaf mule's ear (*W. angustifolia*) has numerous, narrowly spoon-shaped leaves instead. A third species—*W. helenioides*—differs from common mule's ear by its hair-covered leaves. All have equally showy flowers and fragrant roots that were used medicinally by Native Americans.

DODDER FAMILY (CUSCUTACEAE).

Stringy, orangish parasites, which wind around the host plant. Leaves reduced to nonfunctional scales; white flowers in small clusters, with 5 tiny sepals and 5 starlike petals fused into a short tube, 5 stamens, single pistil with superior ovary. Fruit a capsule.

Dodder (*Cuscuta* spp.). P. 301.

These close relatives of the morning glories have altogether lost the ability to photosynthesize and have no green chlorophyll. Soon after seed germination, a wandering stem snakes its way to the nearest host, anchors onto it, penetrates it, and winds around it like so much baling wire. Dense infestations of dodder are conspicuous due to the distinctive orange color of the stems. Fairly specific to their host, dodder sometimes succeeds too well, resulting in the death of the host plant. The small white flowers are produced in short clusters but are seldom eye-catching.

DOGBANE FAMILY (APOCYNACEAE).

Nonwoody perennial plants with opposite, undivided leaves and copious milky sap. Umbellike clusters of small green to pinkish flowers; 5 sepals, 5 separate petals fused into a tube, 5 stamens, two pistils joined together at the top, with 2 separate, superior ovaries. Fruit a follicle.

Indian Hemp (*Apocynum cannabinum*). P. 299.

Indian hemp has stems with some of the best and strongest fibers of any native plant, and it was used extensively as twine to create strong strings and ropes. Limited to moist areas, this species was probably managed by Native Americans by coppicing the stems to create many straight shoots. It has few animal pests, since the milky sap is highly poisonous. The inconspicuous greenish flowers add little color, whereas a sister species (*A. androsaemifolium*) has attractive pink flowers. The latter is rare here. Another member of the dogbane family is periwinkle (*Vinca major*), seen at abandoned homesites. The latter creates a dense, intertwining ground cover with showy trumpetlike blue flowers and comes to us from Europe.

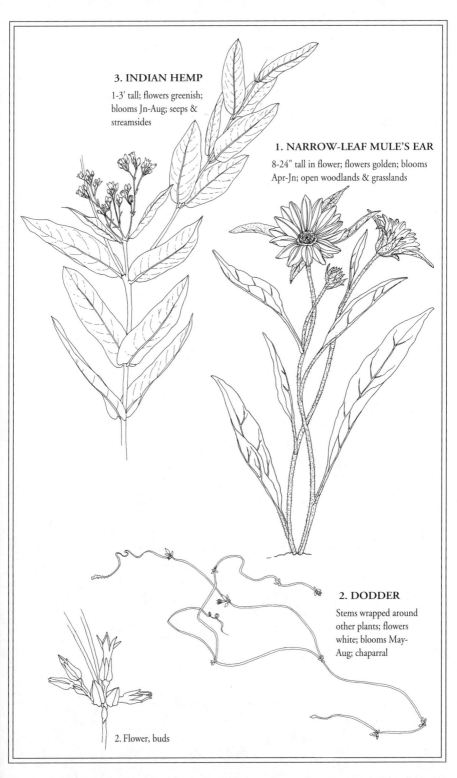

3. INDIAN HEMP

1-3' tall; flowers greenish; blooms Jn-Aug; seeps & streamsides

1. NARROW-LEAF MULE'S EAR

8-24" tall in flower; flowers golden; blooms Apr-Jn; open woodlands & grasslands

2. DODDER

Stems wrapped around other plants; flowers white; blooms May-Aug; chaparral

2. Flower, buds

EVENING PRIMROSE FAMILY (ONAGRACEAE).

Mostly nonwoody plants with undivided, often opposite leaves. Flowers mostly in spikes or racemes; 4 sepals, 4 separate petals attached to floral tube, 4 or 8 stamens, single pistil with inferior 4-chambered ovary. Fruit a capsule.

Boisduvalia (*Boisduvalia* [*Epilobium*] spp.). P. 300.

"Boisduvalia" sounds like a complexly composed technical name, but instead it honors a Frenchman by the name "Bois du Val" (woods of the valley). There is no well-established common name, and few people notice this pretty little annual that is so common in dry open fields at the end of spring. Look closely and you'll see glandular hairs along the leaves, tiny rose-purple flowers with notched petals, and long, tapered seed pods filled with seeds for next year's crop. Unlike its close relative willow herb (*Epilobium* spp.), boisduvalia has no hairs on its seeds, but recent classifications have placed both in the same genus for their highly similar flower design.

Suncups (*Camissonia* spp.). P. 301.

Suncups open their flowers to the midday sun, hence the name. They used to be included in the genus *Oenothera* (evening primrose), but their habit of blooming for day-active pollinators and their knob-shaped stigmas (instead of a four-pointed cross) sets them apart. All suncups prefer open, well-lighted places, whether on coastal dunes and bluffs, inland on grassy hillsides, or on burns after chaparral fires. Some have tiny flowers only a fraction of an inch across, while others—such as golden eggs (*C. ovata*)—have inch-wide flowers borne one after the other. Golden eggs also stands out in its flat rosette of elliptical leaves, the very long floral tube, and the ovary that is actually buried below the soil surface. No one knows, but possibly ants disperse the seeds. Another suncups to look for is the tiny-flowered *C. micrantha*, a sprawling annual with miniature yellow blossoms.

Godetia or Farewell-to-Spring (*Clarkia rubicunda, gracilis, purpurea*). P. 301.

Godetias are also widely known as herald-of-summer or farewell-to-spring, for they bloom at the cusp between spring and summer. Our several species all have delightful cup-shaped flowers with pink or purple petals, sometimes blotched with red or dark purple near the base. Flower size varies, the showiest having striking blossoms over an inch across. Gardeners still call these flowers by their old name, godetia; botanists, however, have classified the group with the clarkias, since there are intermediate forms connecting the fan-shaped flowers of typical clarkias with the cuplike flowers of godetia. The English have long cultivated godetias, but we're just beginning to appreciate these lovely flowers. Easy to grow in the garden, many fine species have horticultural appeal, such as the deep crimson forms of *C. purpurea*, the winecup clarkia. *C. rubicunda*—the ruby cup—has a red blotch at the base of the petals, while *C. gracilis* has no spots and the flower buds nod.

Clarkia (*Clarkia unguiculata* and *concinna*). P. 301.

Clarkias were named to honor Clark of the Lewis and Clark expedition. They are among our most beautiful late-flowering annuals in woodlands, where they light up shaded places. Godetias and clarkias have similar leaves, but the true clarkias have showy, fan-shaped flowers, each petal a single or lobed spatula shape. Elegant clarkia (*C. unguiculata*) has flowering stems to over two feet in a good year, with a long succession of pink-purple fans that are trimmed with red sepals. Red ribbons (*C. concinna*) has intensely rose-pink petals beautifully slashed into

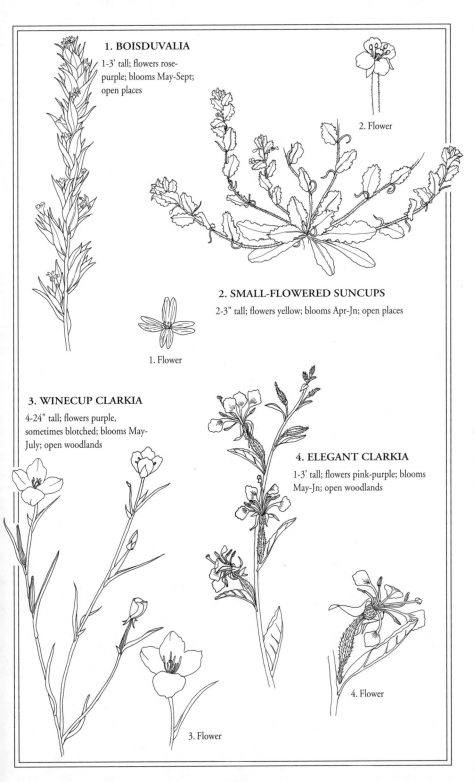

1. BOISDUVALIA

1-3' tall; flowers rose-purple; blooms May-Sept; open places

2. Flower

2. SMALL-FLOWERED SUNCUPS

2-3" tall; flowers yellow; blooms Apr-Jn; open places

1. Flower

3. WINECUP CLARKIA

4-24" tall; flowers purple, sometimes blotched; blooms May-July; open woodlands

4. ELEGANT CLARKIA

1-3' tall; flowers pink-purple; blooms May-Jn; open woodlands

4. Flower

3. Flower

3s, with red, turned-down ribbonlike sepals. These plants seldom reach more than a few inches high and are particularly fond of mossy banks.

Willow Herb (*Epilobium* spp.). P. 302.

Willow herbs look similar to boisduvalias but are seldom glandular, and the seeds bear tufts of hairs for effective wind dispersal. This last feature plus the quick adaptability of the plants make them weedy in their ardor to survive. Willow herbs are so named for the ovate-lance-shaped leaves resembling those of willows, and some are found in the same riparian habitats (for example, *E. ciliatum watsonii,* illustrated here). The tiny pink-purple to purple flowers with notched petals are scarcely noticed because of their small size, but seed pods display conspicuous white-haired seeds. An entirely different habitat is occupied by the summer- to fall-flowering *E. brachycarpum,* with tiny flowers on tall, wandlike stems. It prefers disturbed dry areas. Finally, we have the scaled-down *E. minutum,* a charming miniature annual with pale pinkish flowers on rocky outcrops in mid-spring.

Evening Primrose (*Oenothera* spp.). P. 305.

Evening primroses are among our showiest wildflowers, the individual flowers often measuring three or more inches across. Unlike suncups, flowers open late in the day and remain open through the night, closing soon after the morning sun appears. Their pale yellow or white color assures that they show up against a dark background just when hawkmoths are busy looking for a meal of nectar. The protruding stamens and four-pointed stigma hit the moth on the head while it hovers in place, uncurling its long tongue to deeply probe the long, narrow floral tube. Our most common evening primrose is the yellow-flowered *Oe. elata hookeri* (pictured), an imposing plant growing to several feet high and common around wet places. One of our choicest flowers, however, is the rare Antioch dunes evening primrose (*Oe. deltoides howellii*), occurring only on the sand dunes near Antioch. Here the low-lying plants produce a long season of oversized, fragrant white flowers that fade rose. This oenothera is closely related to the desert birdcage primrose (*Oe. deltoides*), a desert annual with equally showy blossoms.

California or Hummingbird Fuchsia (*Zauschneria californicum* [*Epilobium canum*]). P. 308.

One of the last wildflowers to bloom is our hummingbird fuchsia, which may continue to flower until hard frosts. Out of bloom, the scraggly stems with their pale gray-green leaves are not particularly distinguished, although they do a good job of colonizing rocky embankments, often along stream courses. By contrast the flowers are arresting. Each flower is a long, flared scarlet trumpet, the sepals and petals both colored and the reddish stamens extending far forward beyond the petals. This conformation is perfect for hummingbirds, and reminds us of the design seen in the true fuchsias, which—although they belong to the same family—are in the genus *Fuchsia* and occur in Central and South America. The flower shape and color of zauschnerias differ from those of epilobiums, but some botanists have combined the two in the same genus on other grounds, including the fact that both produce hair-covered seeds. Zauschnerias respond well to the garden, where their exuberance may sometimes get out of bounds but they reward with their late-summer to fall flower show and their attraction of hummingbirds.

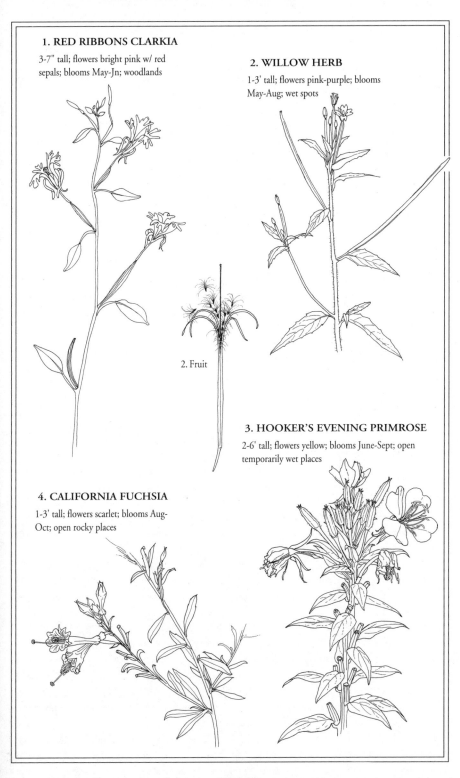

1. RED RIBBONS CLARKIA

3-7" tall; flowers bright pink w/ red sepals; blooms May-Jn; woodlands

2. WILLOW HERB

1-3' tall; flowers pink-purple; blooms May-Aug; wet spots

2. Fruit

3. HOOKER'S EVENING PRIMROSE

2-6' tall; flowers yellow; blooms June-Sept; open temporarily wet places

4. CALIFORNIA FUCHSIA

1-3' tall; flowers scarlet; blooms Aug-Oct; open rocky places

FIGWORT FAMILY (SCROPHULARIACEAE).

Nonwoody plants or small shrubs; leaves variable, often opposite. Flowers usually showy and 2-lipped. Five partly fused sepals; 5 petals arranged as 2 above and 3 below and joined to form a tube or throat; usually 4 stamens in 2 pairs (sometimes 5 or 2); single pistil with superior, 2-chambered ovary; stigma often 2 lobed. Fruit a capsule with numerous seeds.

Wild Snapdragon (*Antirrhinum* spp.). P. 299.

Wild snapdragons are close relatives of the popular annual bedding plants of the same name, but the latter hail from Europe. Our snapdragons are like miniature versions of the garden kind and are mostly annuals, but the plants behave in different fashion. Most native snapdragons are fire followers, germinating in quantity after chaparral has burned, then gradually retreating as shrubs resprout. The most common species (*A. vexillo-calyculatum*), illustrated here, has pale blue-purple flowers, and it climbs by curious prehensile taillike extensions of the flowering stalks that clasp onto nearby vegetation.

Indian Paintbrush (*Castilleja* spp.). P. 301.

Indian paintbrushes are so named because the dense flowering spikes look as though someone dipped them in a bucket of brightly hued paint. On closer inspection the bright reds, oranges, or yellows come mainly from the ends of the floral bracts that underline each flower. Sepals also contribute their color, but the petals are mostly green and are curiously arranged, with the upper two extending out into a long, pointed beak that envelops the stamens and style. This tubular design together with the bright colors makes paintbrushes favorite hummingbird flowers. We have several species; perhaps the easiest to recognize is wooly paintbrush (*C. foliolosa*), whose wooly white leaves make sharp contrast with the host plant to which its roots hook: chamise, with bright green leaves.

Chinese Houses or Blue-eyed Mary (*Collinsia* spp.). P. 301.

The whorls of flowers on our most common collinsia—*C. heterophylla*—are arranged in tiers in spikelike arrangements with a bicolored theme—pale upper lip and dark blue-purple lower lip. This gives the neat and regular appearance of a tiered Chinese pagoda. Chinese houses are annuals, with a propensity for light shade, and they flower toward the end of spring, often in company with clarkias. One of the best identifying features is the middle lower petal—with its pealike keel—that hides between the two outer petals. A similar collinsia with whorled flowers is *C. tinctoria* (tincture plant), whose flowers are palest blue-purple and whose leaves stain fingers. Two collinsias without the tiers of flowers are *C. sparsiflora* and *C. parviflora*, both of which bear single flowers at the tips of short stems. Grown in containers, collinsias continue to bloom over a long period when given periodic water.

Scarlet Monkeyflower (*Mimulus cardinalis*). P. 304.

The monkeyflowers are all noted for their "personality": the two-lipped petals create caricaturelike faces, complete with throat and mouth. In fact the Latin name "mimulus" stands for "little mime." Scarlet monkeyflower is further distinguished by the upper lip protruding to arch over stamens and stigma and by the lower lip sweeping backward. This is to be expected in a hummingbird flower, where the protruding stamens reach out to dust the hummer on the head and the lower lip is recessed so that nectar robbers can't land. Scarlet monkeyflower blooms late in summer and continues into fall, where it lights up streamsides. It is also an easy subject in a natural woodland garden, if given occasional deep waterings.

1. WILD SNAPDRAGON

1-2½ tall; flowers blue-purple; blooms
May-Aug; open chaparral, rocky
places & after burns

2. WOOLY PAINTBRUSH

1-2' tall; flowers red-orange; blooms Mrch-
Jn; edge of chaparral

3. Flower

3. CHINESE HOUSES

8-24" tall; flowers pale & deep
purple; blooms Mrch-Jn; open
woodlands

4. SCARLET MONKEYFLOWER

1-3' tall; flowers scarlet; blooms Jn-Oct;
streamsides

Mickey Mouse Monkeyflower (*Mimulus douglasii*). P. 305.

While the scarlet monkeyflower is a substantial perennial, mickey mouse monkeyflower is a diminutive annual growing only inches high on loose, rocky soils where there's minimal competition with other plants. Most of the plant's energy goes into making those curious flowers, the upper lip resembling a pair of outsize, mickey-mouse-like ears and the lower lip with petals tucked under, so as to appear chinless. These ephemeral annuals have fluctuating numbers from year to year. There are several other annual monkeyflowers, most with rose-purple flowers, but none with the character and noteworthy flower shape that this one has.

Golden Monkeyflower (*Mimulus guttatus*). P. 305.

Another stream lover, golden monkeyflower blooms earlier in the season than the scarlet and is also often found in temporarily wet places. The lank stems grow rapidly as days lengthen, with open racemes of large, butter-yellow snapdragonlike flowers, the lower lip elegantly curved down and sprinkled with brown spots that serve as pollination guides leading into the throat. As in all monkeyflowers, the stigma is two lobed and sensitive to touch, actually closing within seconds after contact. This prevents self-pollination, since bees must force their way inside the throat to locate nectar and upon backing out cannot accidentally dust their new load of pollen on the now-closed stigma. Another stream-dwelling yellow monkeyflower common to the north of us is musk monkeyflower (*M. moschatus*), with slimy leaves and nearly symmetrical light yellow petals. Both are easy subjects in a garden receiving summer water.

Cream Sacs (*Orthocarpus* [*Triphysaria*] *erianthus, and others*). P. 305.

The orthocarpuses are common annuals throughout California's grasslands but don't really come to prominence until their flower spikes open. The narrow, deeply slashed or divided leaves are hidden by surrounding grasses. Each flower is a marvel of design: three bright yellow lower petals inflated like miniature balloons and two upper petals forming a narrow green or red-purple "beak" that extends beyond the lower. The weight of bees on the lower lip bends it down to give access to the shallow nectar tube, while the stamens hidden in the upper lip dust pollen on bees' backs.

Owl's Clover or Escobitas (*Orthocarpus* [*Castilleja*] *purpurascens* [*exserta*] and *attenuatus*). P. 305.

Our two species, which look similar but differ in color, are among our most prolific mid-spring annual wildflowers. *Orthocarpus purpurascens* is a showstopper with its lavish display of pink-purple spikes, while *O. attenuatus* has more modest spikes of white, with only hints of yellow and purple at close range. When *O. purpurascens* mixes with poppies or lupines the color combination is stunning. The dense, moplike clusters of flowers are best served by the common name "escobitas," Spanish for "little brooms." Certainly these flowers are only distantly related to true clovers in the pea family. At close range each flower has three lower petals with a pair of dark spots resembling owl eyes and a protruding beak from the two upper petals. Most of the color comes from the slashed floral bracts whose tips are marked vivid rose-purple, much like their relatives the paintbrushes. One other little-noticed member of the orthocarpus group is *O. pusillus* [*Triphysaria pusilla*], with maroon leaves and diminutive maroon flowers.

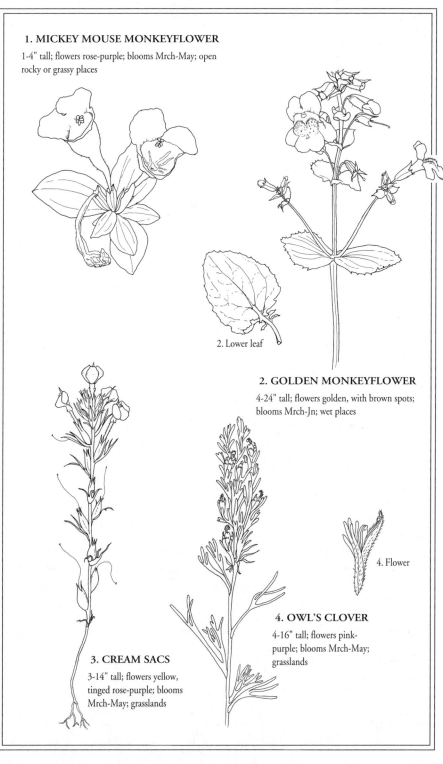

1. MICKEY MOUSE MONKEYFLOWER

1-4" tall; flowers rose-purple; blooms Mrch-May; open rocky or grassy places

2. Lower leaf

2. GOLDEN MONKEYFLOWER

4-24" tall; flowers golden, with brown spots; blooms Mrch-Jn; wet places

4. Flower

4. OWL'S CLOVER

4-16" tall; flowers pink-purple; blooms Mrch-May; grasslands

3. CREAM SACS

3-14" tall; flowers yellow, tinged rose-purple; blooms Mrch-May; grasslands

Bellardia (*Bellardia trixago*).

An abundant European annual, close to orthocarpuses but with unslashed leaves, Bellardia is rapidly gaining ground in north and central California grasslands. Like orthocarpuses, Bellardia has weakly parasitic roots, inclined to join up with grass roots to obtain minerals it is unable to get on its own. Bellardia is attractive, with a folded purple upper lip and broad white lower lip. The flowers are neatly tiered in four precise rows. Sadly, bellardia is purloining habitat from native grassland wildflowers. Another introduced weed in grasslands to our north is *Parentucellia viscosa* (no common name), with hatchetlike upper lip and inflated lower lip, both in bright yellow.

Indian Warrior (*Pedicularis densiflora*). P. 305.

Indian warrior carries on the tradition of its relatives, the paintbrushes and orthocarpuses, by parasitizing plants nearby. Unlike orthocarpuses, however, Indian warrior is particular about its host, preferring such members of the heather family as manzanita and madrone. Being perennial, the colonies enlarge their coverage of ground beneath their hosts from year to year. They're easy to recognize even in early spring as the dark green, sometimes reddish-tinted, fernlike leaves push up. The curious flower spikes bear deep wine-red flowers, with long, narrow hatchet-shaped upper lip and inconspicuous lower lip, and are another enticement to hummingbirds.

Foothill Penstemon (*Penstemon heterophyllus*). P. 305.

Penstemons differ from most figworts in having five stamens; hence the name. Evolution has taken an unusual turn with the fifth stamen, for it is sterile and lacks pollen sacs. In some species this fifth stamen is fashioned into a nectar guide and looks like a hairy tongue sticking out between the two petal lips. Such species are called beard-tongues, but foothill penstemon has an undistinguished fifth stamen hidden within the petal tube. This penstemon, like so many of its relatives, prefers rocky banks along the edges of chaparral, where its inflated yellow flower buds open to varied shades of blue, blue-purple, or pink-purple. Foothill penstemon is easily propagated from tip cuttings and is a stunning short-lived garden flower for the mixed border or rock garden.

Figwort or Bee Plant (*Scrophularia californica*). P. 307.

This is an aggressive, weedy member of the family. The botanical name is derived from the fact that the first European species described was supposed to cure the disease called scrofula. Figwort is a perennial with widely ranging roots, sending up numerous separate purple-tinted stems with triangular leaves. These leaves superficially resemble those of the wood mints (*Stachys* spp.) but lack their characteristic strong odor. The flowers go almost unnoticed, for they're produced on tall, open panicles and are individually small and maroon-red to brownish. Up close the architecture is fascinating, for the upper lip juts out like an awning (probably to keep rain from hurting the pollen sacs underneath), while the lower lip is recessed and has only small lobes. Although dully colored, the flowers are favorite nectar sources for bees.

Common Mullein (*Verbascum thapsus*).

Mullein is a European immigrant with great tenacity for roadsides and newly exposed roadbanks. As a biennial it produces conspicuous wooly-leafed rosettes the first year, then sends up a several-foot-tall flowering spike the second. These spikes carry a long succession of open, cup-shaped yellow flowers in late spring and summer. The plants die after seeds have ripened.

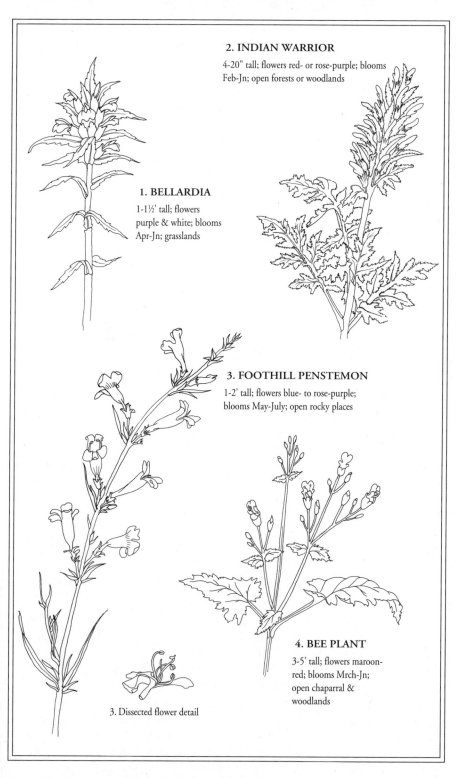

2. INDIAN WARRIOR

4-20" tall; flowers red- or rose-purple; blooms
Feb-Jn; open forests or woodlands

1. BELLARDIA

1-1½' tall; flowers
purple & white; blooms
Apr-Jn; grasslands

3. FOOTHILL PENSTEMON

1-2' tall; flowers blue- to rose-purple;
blooms May-July; open rocky places

4. BEE PLANT

3-5' tall; flowers maroon-
red; blooms Mrch-Jn;
open chaparral &
woodlands

3. Dissected flower detail

A second mullein (*V. blattaria*, moth mullein) is also making inroads into natural areas. It has narrow, more modest spikes of wide-open white or yellow blossoms.

Speedwell (*Veronica americana*). P. 308.

The veronicas are well known throughout the world, and they include weedy annuals in lawns as well as this native brook follower. Our native species is entirely restricted to wet places, where the stems creep, sending down roots as they wander around. Tiny flowers are produced in long succession and are charming up close. The petals have been reduced to four and spread widely with only a very short tube, so that the two-lipped trait so well established in this family is not evident here. The stamens are also reduced to two. Petals are flushed with darker lines—nectar guides—and the entrance to the tube may be marked with yellow, another clue to the location of nectar. Veronicas are close relatives to the shrubby hebes we cultivate from New Zealand.

FLAX FAMILY (LINACEAE).

Nonwoody plants with narrow, undivided leaves. Flowers borne on open branches; 5 sepals and 5 separate petals that fall easily, 10 stamens, single pistil with superior ovary, and 3 to 5 separate styles. Fruit a capsule.

Native Flax (*Hesperolinon* spp.). P. 303.

Our native flaxes are such tiny annuals, they're scarcely noticed even in full flower. Once you know the delicate, airy appearance of the plants, you can expect to find them on blue serpentine rock or dark lava from mid- to late spring. The ephemeral flowers are yellow in *H. breweri* (pictured) or white to pale purplish in others. Many of the species have a restricted distribution and are good indicators of specialized habitats.

Common Flax (*Linum usitatissimum*).

Common flax is the source of commercial flax fiber from which linen is made (note the similarity between the words "linum" and "linen"). European in origin, it came to its new homeland by accident and has found the climate so much to its liking, it appears native. Look for it in grasslands in the fog belt, where its delicate pale purple flowers open in morning but soon close and fade.

FUMITORY FAMILY (FUMARIACEAE).

Nonwoody plants, often from creeping rootstocks, leaves highly divided, often fernlike. Irregular heart-shaped flowers; 2 sepals, 4 petals arranged in irregular pairs, 6 stamens hidden inside petals and fused into 2 clusters, single pistil with superior ovary. Fruit a capsule.

Golden Eardrops (*Dicentra chrysantha*). P. 302.

Golden eardrops should be renamed golden hearts, for the myriad flowers borne on six-foot stalks look like upright yellow hearts. This short-lived perennial is rare on Mt. Diablo, appearing in great numbers after chaparral fires and then slowly dying out, as tall shrubs crowd and shade it out. After the 1977 fire, golden eardrops appeared by the thousands. The basal tufts of bluish green ferny leaves identify it out of flower.

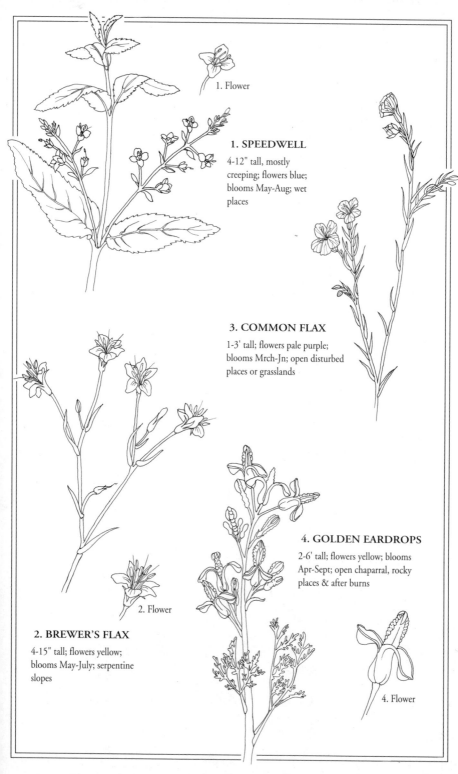

1. Flower

1. SPEEDWELL

4-12" tall, mostly
creeping; flowers blue;
blooms May-Aug; wet
places

3. COMMON FLAX

1-3' tall; flowers pale purple;
blooms Mrch-Jn; open disturbed
places or grasslands

4. GOLDEN EARDROPS

2-6' tall; flowers yellow; blooms
Apr-Sept; open chaparral, rocky
places & after burns

2. Flower

2. BREWER'S FLAX

4-15" tall; flowers yellow;
blooms May-July; serpentine
slopes

4. Flower

Western Bleeding Heart (*Dicentra formosa*). P. 302.

Western bleeding heart is common in northern California, but in our area is restricted to a few locales. It seeks moist, shaded forests and is well represented throughout most of the redwood belt. There, its creeping roots soon establish large colonies of coarse ferny leaves. The lovely flower clusters are lifted above the leaves in mid- to late spring, with nodding rose-purple to pale pink, upside-down hearts. Seeds bear curious white oil bodies and are responsible for this species being ant dispersed; ants carry off the shiny black seeds, nibble off the oil bodies, and discard the rest, leaving the main seed intact to germinate. Western bleeding heart is a charming ground cover for a shaded garden but, with extra water, becomes invasive.

GENTIAN FAMILY (GENTIANACEAE).

Nonwoody plants with opposite, undivided leaves. Showy flowers; 4 or 5 fused sepals, 4 or 5 fused petals folded in bud, 4 or 5 stamens fused to base of petal tube, single pistil with superior, partly 2-chambered ovary. Fruit a capsule.

Canchalagua (*Centaurium* spp.). P. 301.

Native gentians are rare with us; our most typical genus is this group of annual wildflowers that appear in late spring and early summer. Although they have elliptical, gentianlike leaves in neat pairs, the bright rosy or shocking pink flowers come as a surprise. Up close, these flowers are marvels of design, with strangely twisted anthers that release pollen as they unwind. Look for canchalagua in grassy spots that have been moist. Another uncommon gentian relative is a tiny annual called *Cicendia quadrangularis* that lives on the edges of vernal pools and has bright yellow flowers.

GERANIUM FAMILY (GERANIACEAE).

Nonwoody plants with lobed to compound leaves that have pairs of stipules at the base. Flowers borne singly or in umbels, usually pink or purple; 5 sepals and 5 separate petals, 5 or 10 stamens, single pistil with 5-chambered superior ovary and beaklike styles elongating in fruit.

Filaree or Clocks (*Erodium* spp.).

Erodiums differ from geraniums by their pinnately lobed to compound leaves, some suggesting filagree (an approximation of the common name). The name clocks comes from the long, pointed styles atop the fruit, which separate suddenly into two hands at different angles like those on a clock face. These styles remain attached to a section of the ovary with the seeds intact, and they ride on human clothing or animal fur, effectively dispersing these widespread weeds. All of our filarees are from the Old World and may have been brought in as fodder plants for cattle. Now they color vast grasslands and deserts with a purple haze in early spring.

Wild Geranium or Storksbill (*Geranium* spp.).

Most of our wild geraniums are also introduced—either from eastern North America or Europe—and are common garden or grassland weeds. Unlike garden geraniums (most of which are really in the genus *Pelargonium*), wild geraniums have tiny flowers and leaves that are shallowly to deeply palmately lobed (resembling mallow leaves). In fruit the ovary is crowned with the same elongated beak as is found in erodiums, but in most cases the elastic style sections suddenly curl back, forcibly heaving out the seeds.

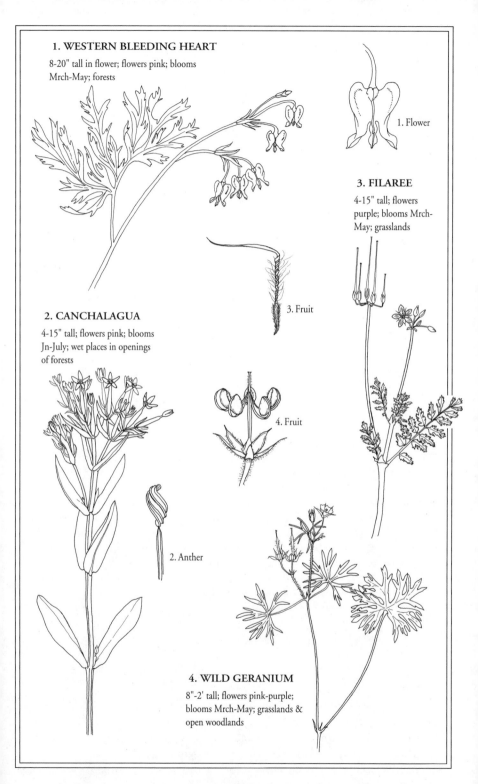

1. WESTERN BLEEDING HEART

8-20" tall in flower; flowers pink; blooms
Mrch-May; forests

1. Flower

3. FILAREE

4-15" tall; flowers
purple; blooms Mrch-
May; grasslands

3. Fruit

2. CANCHALAGUA

4-15" tall; flowers pink; blooms
Jn-July; wet places in openings
of forests

4. Fruit

2. Anther

4. WILD GERANIUM

8"-2' tall; flowers pink-purple;
blooms Mrch-May; grasslands &
open woodlands

IRIS FAMILY (IRIDACEAE).

Nonwoody plants from rhizomes, sword-shaped leaves arranged in flat sprays and overlapping at the base. Showy flowers are protected inside floral bracts before opening. Three colored sepals and 3 colored petals, 3 stamens, single pistil with 3-chambered inferior ovary. Fruit a capsule.

Douglas Iris (*Iris douglasiana*). P. 303.

Douglas iris is our only reasonably common wild iris. Growing from shallow, horizontally creeping rhizomes, each plant grows in circular fashion, thereby increasing in diameter every year. Douglas iris is one of our few wildflowers that overgrazing hasn't affected; in fact, Douglas iris benefits because of its poisonous leaves. The glossy, dark green leaves are apparent all year, but the flowers appear anywhere from the new year through the middle of spring according to temperatures. Each flower is like a delicate orchid whose colors range from palest lavender (occasionally pure white) to deep, rich blue. The three sepals (called falls) actually are showier than the petals and are turned down at their ends; they wear a delicate network of contrasting yellow or white veins that show the way to the nectar tube. Above each sepal sits a stamen and above that (and hidden by it) a petallike style branch. If you look just below the cleft in this style branch, you'll find the minute flap of tissue that functions as the stigma. As a bee crawls down the sepal, it first hits the stigma and rubs off pollen from a previous flower, then farther on it picks up a new load of pollen. Now as it backs out, the flower cannot be accidentally self-pollinated, for the stigma moves in one direction only! Douglas iris is one of our most easily adapted wildflowers for the garden; many nurseries carry it or hybrids between it and other native irises, with a wide range of colors: yellow, bronze, red-purple, blue, lilac, violet, and white.

Blue- and Yellow-Eyed Grasses (*Sisyrinchium bellum* and *californicum*). P. 307.

These iris relatives are called grasses because of their grasslike leaves, but in reality the leaves look much more like miniature versions of true irises than grasses. In fact, everything about them is like a scaled down iris plant except for the flowers. The floral bracts hide many flowers, only one of which emerges at a time. The sepals and petals look much alike, forming a flattened saucer in the middle of which is a short nectar tube. Blue-eyed grass actually has blue-purple flowers with darker stripes and a yellow center or eye, and it is common throughout our grasslands. Yellow-eyed grass is particular about its habitat, favoring wet seeps, and has uniformly golden-yellow blossoms. Its leaves have the curious quality of turning black as they wither. Both species are easily naturalized in a meadow or wild garden.

LILY FAMILY (LILIACEAE).

Nonwoody plants from bulbs, corms, or tubers; leaves parallel veined, sometimes grasslike. Flowers arranged in racemes, panicles, or umbels; 3 sepals and 3 petals often both colored and similar in shape, 3 or 6 stamens, single pistil with 3-chambered, superior ovary, 3-lobed stigma. Fruit a berry or capsule.

Wild Onions (*Allium* spp.). P. 299.

Alliums occur across the northern hemisphere, but the kinds used for food or flavoring—leeks, garlic, chives—originated in the Old World. California's wild onions are instantly recognized when a flower, leaf, or stem is crushed, and all are edible. It would be senseless to uproot these plants for a meal when their beauty contributes so much to rocky, grassy slopes in spring. The species are difficult to identify. Books often refer to the pattern on the skin of the bulbs, though

BLUE-EYED GRASS

4"-2' tall; flowers blue-purple w/ yellow center; blooms Mrch-May; grasslands

DOUGLAS IRIS

6-30" tall; flowers blue-purple; blooms Mrch-May; open forests

SICKLE-LEAF ONION

1-4" tall; flowers pink- to rose-purple; blooms Apr-Jn; loose rock scree

digging one up to check this is a sure way to kill the plant. Species in our area include: *A. crispum*, wavy onion, with wavy or "crisped" inner petals and lovely rose-purple color; *A. amplectens*, white lava onion, with pure white to pale pink flowers; *A. falcifolium*, sickle-leaf onion, with flattened stem, sickle-shaped leaves, and purple flowers; *A. acuminatum*, Hooker's onion, with two to three leaves and bright rose flowers that have petals inflated at the base; *A. unifolium*, single-leaf onion, with showy umbels of pale pink starlike flowers and *A. serra*, pink onion, with cylindrical leaves and plain, pink flowers. Single-leaf onion favors moist grasslands or the edge of woodlands, while the others are to be looked for on rocky scree in full sun or semi-shade. Wild onions, particularly *A. unifolium*, naturalize easily in the garden but are eaten by gophers.

Harvest Lily or Brodiaea (*Brodiaea elegans*). P. 300.

Much confusion surrounds the naming of the brodiaeas and their relatives, and the only common true brodiaea in our area is the harvest lily. This pretty corm-bearing plant waits until the grasses have burned brown before flowers swell and open. The open umbels of blue-purple flowers are particularly welcome when other flowers have already passed, and often find company with mariposa tulips. The true brodiaeas have shiny flowers with a gloss that makes them look waxy; they also have only three functional stamens. The sterile stamens look like an extra row of internal, white petals, easily seen against the blue of the true petals. Brodiaeas furnished one of the reliable sources of "Indian potatoes," a broad category for many of California's edible bulbs. See *Triteleia* and *Dichelostemma*, below, for other brodiaeas.

Fairy Lanterns or Globe Tulips (*Calochortus albus* and *pulchellus*). P. 300.

Calochortus is Greek for beautiful grass, and although not a grass at all, these species have grasslike leaves and truly beautiful flowers. As described below, mariposas have tulip-shaped flowers while star tulips have shallow, saucer-shaped petals. The fairy lanterns, in contrast, are unique for their nodding, globe-shaped flowers whose petals overlap to enclose and hide the nectaries, stamens, and pistil. We're fortunate to have two fairy lantern species in our area. *C. pulchellus*, the Mt. Diablo globe tulip, is restricted to Mt. Diablo itself and has lovely lemon yellow flowers. Although closely related to Diogene's lantern (*C. amabilis*), which grows north of the Golden Gate, its shape differs. Our other species is the white globe tulip (*C. albus*), characteristic of the Livermore Hills and southern end of our territory. Although the petals are white, the sepals and base of the petals are often flushed with pink, making this one of our most exquisite wildflowers.

Mariposa Tulips (*Calochortus* spp.). P. 300.

Our second group of calochortuses has a totally different flower shape: upright, vase- or cup-shaped blossoms reminiscent of wild tulips. Unlike garden tulips, each petal bears a nectar gland decorated with hairs and often blotched, streaked, and pencilled with fantastic patterns in brown, mahogany, red, black, or yellow. These intricate patterns are reminiscent of butterfly wings, as the Spanish name "mariposa" reminds us. Mariposas appear at the very end of spring when grasses have dried or gone dormant, and they are all the more welcome then. Strangely, the beauty of these flowers seems unappreciated by their insect guests, for they're most often visited by tiny beetles. Our three species include *C. luteus*, yellow mariposa, with bright yellow petals and spots or lines of darkest brown; *C. venustus*, the "white" mariposa, although the flowers are often flushed pink or purplish and marked with kaleidoscopic variety of splotches, spots, and streaks; and *C. splendens*, the lilac mariposa, with purple stamens and lilac petals. The last is our most restricted species, appearing only along the south side of Mt. Diablo (and is common in southern California).

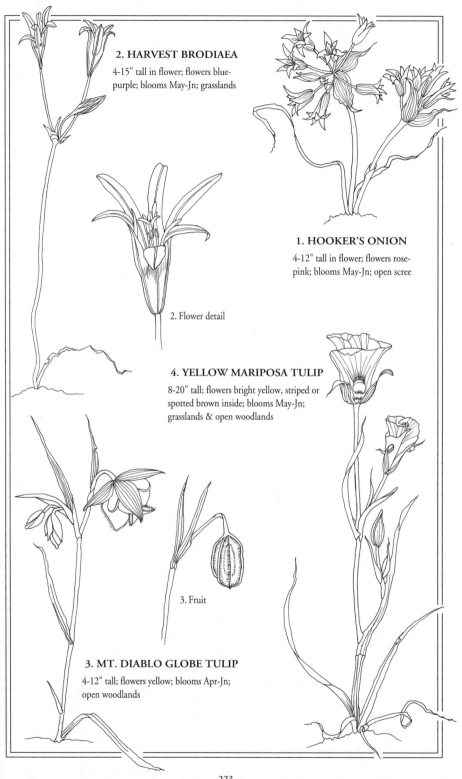

2. HARVEST BRODIAEA

4-15" tall in flower; flowers blue-purple; blooms May-Jn; grasslands

1. HOOKER'S ONION

4-12" tall in flower; flowers rose-pink; blooms May-Jn; open scree

2. Flower detail

4. YELLOW MARIPOSA TULIP

8-20" tall; flowers bright yellow, striped or spotted brown inside; blooms May-Jn; grasslands & open woodlands

3. Fruit

3. MT. DIABLO GLOBE TULIP

4-12" tall; flowers yellow; blooms Apr-Jn; open woodlands

Oakland Star Tulip (*Calochortus umbellatus*). P. 300.

Oakland star tulip is a diminutive bulb, with modest, shallow, saucer-shaped white or pale purple flowers. Much less showy than the globe tulips or mariposas, it is still a charming flower, but you have to search in just the right place and at the right time to see it. Oakland star tulip often indicates serpentine soils and, although first discovered in the Oakland hills, is more abundant on the serpentines of Marin County and only occasional on Mt. Diablo.

Soap Plant or Amole (*Chlorogalum pomeridianum*). P. 301.

Soap plant is perhaps our most abundant large-bulb plant, with bulbs reaching the size of an overgrown onion. You first notice it in winter when the wavy, bluish green leaves reemerge from the earth, but you have to wait until spring's end for the bloom, and even then you'll miss it unless you're out after four p.m. In fact the species name stands for post meridian in allusion to the afternoon- and evening-blooming flowers. Then the thin, spidery white petals look like they're suspended in air because of the open disposition of the flowers on thin stalks. By morning, flowers are tightly shut—never to open again—but there's a long succession of flower buds. Not only were the bulbs used for soap by the Native Americans but the fibers around the bulb made brushes, the gummy substance inside the bulb was glue, and the whole bulb could be crushed and thrown into a pond or slow stream to stupefy fish, or baked in an earth oven as food. Soap plant is also our easiest native bulb to naturalize in the garden.

Blue Dicks and Ookow (*Dichelostemma capitatum* and *congestum*). P. 302.

Blue dicks and its sister species, ookow, are most often referred to as brodiaeas. Unlike harvest brodiaea, however, their flowers are not waxy and are arranged in dense clusters. Flowers bear a crown of "appendages" around the stamens. (The genus name comes from "dichelos," meaning toothed, and "stemma," for crown, describing these appendages.) Blue dicks is as abundant as soap plant, but it flowers starting in early spring. The slender flowering stalks are often prettily curved as they grow, bearing at the top headlike clusters of pale blue flowers surrounded by deep purple floral bracts. The edible bulbs were favorite Indian potatoes. Ookow is much less common—scattered through grasslands and woodlands—and blooms several weeks later. At first inspection it's hard to distinguish ookow from blue dicks, but ookow's flowers are arranged in compact racemes, lack the purple bracts, and have three stamens instead of blue dicks' six.

Fairy Bells (*Disporum hookeri*). P. 302.

The modest fairy bells pushes up asparaguslike shoots in early spring in the shade of moist forests. Watch it carefully, for the pale green bells hang beneath the leaves just short of the stems' ends. The showiest part is the bright orange, rounded-oblong fruits that ripen in summer. Fairy bells have leaves much like those of our false Solomon's seals (*Smilacina* spp.), but the stems branch, and the flowers and fruits are altogether different.

Checker Lily or Mission Bells (*Fritillaria* spp.). P. 303.

These exquisite lilies seldom occur with the abandon of so many of our spring bulbs but always remain special. By far the most common species is the checker lily (*F. lanceolata* [*affinis*]), with its hanging belllike flowers decorated with a checkerboard pattern of browns, greens, purples, and white, and marked by yellow glands inside. The whorled leaves are the first evidence of this plant in early spring. Many color variations and size differences occur in this shade-loving species. White fritillary (*F. liliacea*) is among our truly rare plants, occurring sporadically in the

1. OAKLAND STAR TULIP

3-10" tall; flowers white or pale purple; blooms Mrch-May; serpentine slopes

2. SOAP PLANT

2-6' tall in flower; flowers whitish; blooms May-July; open woodlands & grasslands

2. Bulb

3. Flower detail

5. CHECKER LILY

1-3' tall; flowers mottled brown/ green/white; blooms Mrch-May; forests & woodlands

3. BLUE DICKS

1-2½' tall in flower; flowers blue-purple; blooms Mrch-May; grasslands & open woodlands

5. Fruit

4, FAIRY BELLS

1-2½' tall; flowers greenish; blooms Mrch-May; forests

Berkeley and other East Bay hills, usually in open serpentine grasslands. The white bells are trimmed with green midveins and glands. Fritillarias often take a year "off" from flowering; then they send up only a single broad leaf.

False Solomon's Seal (*Smilacina* spp.). P. 307.

Our two false Solomon's seals emerge from forest soils at winter's end with fairy bells. Starry false Solomon's seal (*S. stellata*) has slender racemes of tiny, six-pointed white stars; fat false Solomon's seal (*S. racemosa*) has congested panicles of minute white flowers with sturdier leaves and stems. The latter's flowers smell like lily-of-the-valley at close range. Flowers are followed by red-speckled berries. The former has no discernible odor and purplish berries with darker stripes. Other differences include the wandering tendencies of slender false Solomon's seal's roots, which soon grow to form large colonies in the shade of redwoods; fat false Solomon's seal forms tight, slowly-expanding clumps.

Trillium or Wake-robin (*Trillium* spp.). P. 307.

"Trillium" is simply Latin for "parts in threes" and alludes to the fact that not only the petals, sepals, and stamens are based on the number three but also there are only three leaves. These leaves are curious for a lily, as their vein pattern forms a distinct interjoined network, while most other lily relatives have leaves with parallel veins. The three broad leaves are arrayed in a whorl atop a stem that emerges in late winter, just in time for the flowers to open in early spring. Trilliums produce but one flower per plant, and our two species are easily identified by them: *T. ovatum* (coast trillium) has a flower raised on a slender stalk above the leaves and opens pure white and fragrant, then slowly fades to deep rose (a sign to pollinators that the nectar has dried up); *T. chloropetalum* (giant trillium) has a flower sitting smack in the middle of the leaves. *T. chloropetalum* may have white flowers on some plants, maroon-red on others, and even occasionally green (the origin of the species name "chloropetalum"). Coast trillium is common in redwood forests; giant trillium is occasional in mixed hardwood forests. Trilliums are handsome subjects for the shade garden but should never be dug from the wild. They can be grown from seed, though it takes eight or more years to reach flowering size.

Ithuriel's Spear and White Brodiaea (*Triteleia laxa* and *hyacinthina*). P. 308.

Our third group of brodiaeas is characterized by open umbels of nonwaxy flowers with six fertile stamens and no crown of appendages. Triteleias prefer open woodlands or grasslands and are the most varied group for flower color. Ithuriel's spear has deep blue-purple to sky blue funnel-shaped flowers (depending on locale) and in years of abundant rains may color fields by the thousands. Its edible corms were eaten raw or cooked. White brodiaea may be equally abundant but is usually found in temporarily wet meadows and has white, shallowly bowl-shaped flowers, sometimes flushed purple outside. Other triteleias are rare in our area, but many additional kinds occur elsewhere, including several with yellow flowers.

Star Lily or Death Camas (*Zigadenus fremontii*). P. 308.

All zigadenes have poisonous bulbs, and they were sometimes dug and eaten by mistake by Native Americans before bloom. Out of flower, they resemble the highly esteemed, edible blue-flowered camas (*Camassia* spp.), which is missing from our area. Zigadenes have basal clusters of shiny, straplike leaves, followed by long-flowering racemes of creamy, star-shaped flowers that are marked at the base with a pair of yellow-green glands. These glands are claimed to look like oxen yokes, which is what the genus name means. Fremont star lily is common in open woodlands in mid-spring and is the only reasonably common species in our area.

GIANT TRILLIUM

1-2' tall; flowers maroon, pink,
or white; blooms Apr-May;
forests or dense woodlands

FAT FALSE SOLOMON'S SEAL

1-2½' tall; flowers cream color; blooms
Mrch-May; forests

ITHURIEL'S SPEAR

6-30" tall; flowers blue or
blue-purple; blooms Apr-Jn;
open woodlands

FREMONT STAR LILY

1-3' tall; flowers cream color with
greenish center; blooms Mrch-
May; open woodlands

LIVE FOREVER FAMILY (CRASSULACEAE).

Nonwoody plants with fleshy, water-storing, undivided leaves arranged in rosettes. Flowers often in open cymes; 5 sepals and 5 separate petals, 5 or 10 stamens, 5 partly fused to separate pistils with superior ovaries. Fruit a capsule.

Pygmy Weed (*Crassula* spp.). P. 301.

Pygmy weed is well named, for it is the world's tiniest succulent. Often the whole plant measures no more than an inch high, but everything is perfectly miniaturized: tiny, beadlike, fleshy leaves and equally minute greenish flowers (which you have to look carefully with a hand lens to see). Owing to its size, pygmy weed is a poor competitor and ekes out a living in truly unusual places: hard-packed soils that temporarily hold water. (These are seldom big enough to qualify as vernal pools.) Pygmy weed is most noticeable when the leaves have turned bright red, an indication that extra sugars are being converted to protective red pigments.

Hot Rock Dudleya (*Dudleya cymosa*). P. 307.

Dudleyas have narrow to broad lance-shaped leaves arranged in close rosettes next to the soil, usually seated against rocks. The deeply probing lower stem and roots anchor plants to precarious perches and serve to delve under rocks for hidden moisture. Hot rock dudleya is so called because it grows on bare rock faces that sizzle in summer and fall. Often by fall's end many leaves have shriveled, indicating the gradual withdrawal of water during the long summer drought. Hot rock dudleyas light up their environment with the vivid red-orange to yellow-orange blossoms, although pale yellow flowers are more common in our area.

Common Stonecrop (*Sedum spathulifolium*). P. 307.

Sedums differ from dudleyas in the way they produce their flowers: dudleyas' flowering branches spring from between leaves; sedums bear the flowering stalks directly from the middle of the leaf rosette. This is our most common sedum, partial to mossy, lightly shaded rocky places, where there isn't much competition. Companions include rock ferns and various saxifrages, and red larkspur. The species name refers to the spatula-shaped leaves, which in this species may be bluish green, green, or gray. The bright yellow, starlike flowers add drama in late spring.

MADDER OR COFFEE FAMILY (RUBIACEAE).

Plants with opposite or whorled leaves, sometimes with pairs of stipules. Ours have tiny starlike flowers; 4 or 5 minute sepals, 4 or 5 separate petals (with short tube), 4 or 5 stamens, single pistil with inferior, often 2-part ovary. Fruit a pair of nutlets.

Bedstraw (*Galium* spp.). P. 303.

Bedstraws are seldom noticed unless they accidentally cling to clothing or fur. Yet once seen they're easy to remember, for no other plants combine whorled leaves (three to six or more per node), square stems, and minute, recurved prickles that cause stems to clasp nearby objects. The several species are another matter: some are nonnative weeds, while others occur naturally. You have to look twice to see if they're in flower; the flowers are minute four- or five-pointed green to whitish stars (the perfect size to see by hand lens). Two distinctive native species are *G. porrigens* (pictured), with long, climbing stems; and *G. andrewsii*, with unusual sharp, prickly, needlelike leaves. The former species is distinctive in having separate male and female plants (and so being dioecious). The aggressive alien, *G. aparine*, is abundant in disturbed sites.

2. HOT ROCK DUDLEYA

2-10" tall in flower; flowers pale
yellow; blooms May-Jn; rocky places

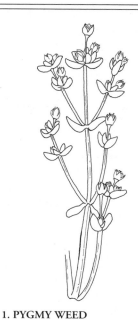

1. PYGMY WEED

1-3" tall; flowers greenish; blooms Feb-May;
open temporarily moist spots

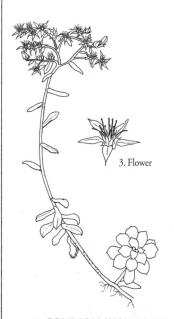

3. Flower

3. COMMON STONECROP

4-12" tall in flower; flowers bright yellow;
blooms May-July; rocky slopes

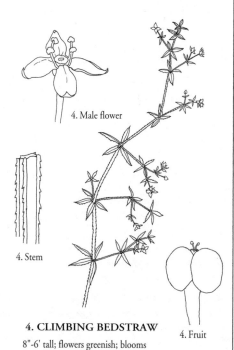

4. Male flower

4. Stem

4. CLIMBING BEDSTRAW

8"-6' tall; flowers greenish; blooms
Mrch-Jn; woodlands & chaparral

4. Fruit

Field Madder (*Sherardia arvensis*).

This diminutive wildflower comes to us from Europe yet is perfectly at home in our open grassy fields. Here the flowers may color areas with pale blue-purple stars in mid-spring. In other details, field madder looks much like the bedstraws, with whorled leaves and star-like flowers.

MALLOW FAMILY (MALVACEAE).

Nonwoody plants or small shrubs with broad, mostly palmately veined leaves, often also with starlike hairs (see with a good hand lens). Flowers hollyhocklike; 5 sepals and 5 floral bractlets resembling sepals, 5 separate petals, numerous stamens fused by their filaments into a hollow tube, single pistil with many-chambered superior ovary. Fruit consists of schizocarps: sections that separate like a sliced cheese wheel.

Mallow or Cheeses (*Malva* spp.).

Mallow is among our most common weeds, especially in gardens, flourishing best where soils are high in nitrogen. The nearly round leaves are fastened to the top of a strong taproot and stems elongate to bear a long succession of miniature mallow- or hollyhocklike blossoms of pale purple. The several species look closely similar, and all are readily identified in fruit by the green, cheese-wheel-like ovary, which darkens to purple. The "cheeses" are sometimes nibbled as snacks, and the seeds are carried by a gluey substance when wetted.

Checker Bloom (*Sidalcea malvaeflora*). P. 307.

Checker bloom announces itself with bright pink hollyhock blossoms by mid-spring. These are irresistible to a wide variety of insects, assuring efficient pollination, but checker bloom throws in an unusual twist to promote cross-pollination: some flowers have stamens and pistils; others have pistils only. Another noteworthy feature in sidalceas is the change in leaf shape from the base of the plant up the flowering stems; basal leaves are shallowly notched; upper leaves are deeply slashed or lobed.

MILKWEED FAMILY (ASCLEPIADACEAE).

Nonwoody perennials with opposite or whorled, undivided leaves and copious milky juice. Flowers in umbels: 5 turned-back sepals and 5 separate turned-down petals; stamens, style, and stigma fused into a central "gynostegium" with a row of 5 nectar cups (hoods) around its periphery; 2 separate superior ovaries. Fruit a follicle; seeds with long hairs.

Milkweeds (*Asclepias* spp.). P. 300.

Milkweeds are arresting, unusual plants and were used by Native Americans for the strong fibers in the stems (second in quality only to Indian hemp). Although the milky juice is poisonous and wards off most munching or browsing animals, the caterpillars of the Monarch butterfly feed exclusively on milkweed parts, storing the poisons to make themselves vile tasting to birds. Few birds that have tasted these caterpillars return for more. Milkweeds are also noteworthy for their ornately complex pollination machinery. Instead of separate stamens, the stamens are fused to the style, but the filaments of these stamens are enlarged into colorful, nectar-bearing cups. The cups themselves sometimes also bear protruding "horns" (which probably act as tongue guides for pollinators). The anther part of the stamen makes a fused ball of pollen attached to a thread with a sticky end; when a pollinator's leg is positioned just right, the sticky end attaches itself with thread and pollen ball intact. At a visit to another flower, the

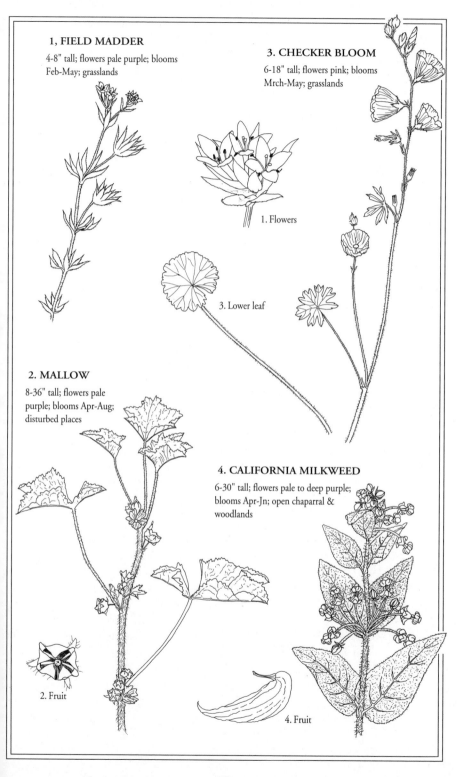

1, FIELD MADDER

4-8" tall; flowers pale purple; blooms
Feb-May; grasslands

3. CHECKER BLOOM

6-18" tall; flowers pink; blooms
Mrch-May; grasslands

1. Flowers

3. Lower leaf

2. MALLOW

8-36" tall; flowers pale
purple; blooms Apr-Aug;
disturbed places

4. CALIFORNIA MILKWEED

6-30" tall; flowers pale to deep purple;
blooms Apr-Jn; open chaparral &
woodlands

2. Fruit

4. Fruit

pollinator's leg may chance to rub against one of the five vertical stigmatic slits, and the pollen ball is inserted. Each species has its own special shape for pollen ball and stigma, just like a lock and key. Our two milkweeds are *A. californica* (California milkweed), with white wooly leaves and red-purple flowers—a knockout combination—and *A. fascicularis* (whorled milkweed), a slender plant with whorled narrow green leaves and small greenish or purplish white flowers.

MINT FAMILY (LAMIACEAE).

Aromatic nonwoody plants or small shrubs, with opposite leaves on square stems. Small flowers are borne mostly in whorls or tight heads and are irregular and 2-lipped; 5 fused sepals (may be hard to count), 5 fused petals—2 upper and 3 lower, fused to form a tube—2 or 4 stamens, single pistil with superior, 4-lobed ovary. Fruit 4, 1-seeded nutlets.

Horehound (*Marrubium vulgare*).

Horehound is familiar to those who remember the horehound drops used for throat problems. Its herbal properties were the reason it was first brought to the New World from the Old. At first grown in gardens, it has escaped and now grows contentedly along dry, rocky stream courses or in old pastures. The quilted leaves are typically mintlike, although fresh they are not particularly strongly scented; the white flowers are small and in dense ball-like whorls up the stem.

Mints (*Mentha* spp.).

The familiar mints used for flavoring candy and mint juleps come to us from Europe but have long been grown in California. Some have made their way from gardens into the natural environment, where they often outcompete natives; most mints have aggressively spreading underground roots. Probably the reason they haven't become even more widespread is that they're partial to permanently moist places, and so only favor disturbed streams and vernal pools. The most familiar of these mints is *M. spicata*—spearmint—with bright green, wrinkled leaves and pale purple flowers. Also common is the intensely mint scented European penny-royal (*M. pulegium*), with near-round leaves and attractive spires of pale purple flowers in neat tiers. Our only native mentha is *M. arvensis* or field mint, whose appearance is occasional on the edge of freshwater marshes. Its small flowers are white or palest purple and the leaves are not notably wrinkled but carry the unmistakable odor of mint.

Coyote Mint (*Monardella villosa*). P. 305.

Coyote mint is sometimes referred to as western pennyroyal, but by that name it is confused with the true pennyroyal. Simply referred to as monardellas, they are either delicate annuals or, more commonly, bushy, low-growing shrublets of dry, rocky places, often on the edge of chaparral. The strongly scented leaves immediately identify them, although the smell is less that of pure mint than a mixture with overtones of sage. A pleasantly flavored herbal tea can be brewed from the leaves. Or the plants can be appreciated for their loveliness alone. The densely packed heads of near-white to purple flowers have flowers with a tubular design that assures that only pollinators with long tongues—bumblebees and butterflies—are successful. Coyote mints also deserve a place in the drought-adapted garden, where they flower well in summer.

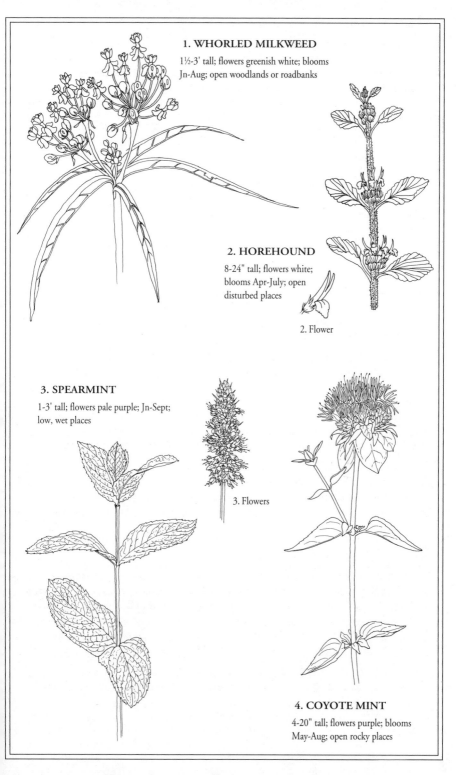

1. WHORLED MILKWEED

1½-3' tall; flowers greenish white; blooms
Jn-Aug; open woodlands or roadbanks

2. HOREHOUND

8-24" tall; flowers white;
blooms Apr-July; open
disturbed places

2. Flower

3. SPEARMINT

1-3' tall; flowers pale purple; Jn-Sept;
low, wet places

3. Flowers

4. COYOTE MINT

4-20" tall; flowers purple; blooms
May-Aug; open rocky places

Vernal Pool Mint (*Pogogyne serpylloides*) P. 306.

Where vernal pools haven't been trashed by cows wandering around or by the trampling of horses, the European pennyroyal is replaced by a true native: vernal pool mint. These tiny annuals are immediately apparent if you walk over them, and the released mint odor is as strong as that of true pennyroyal. To see vernal pool mints in flower requires patience, for the water completely evaporates before the tiny purple flowers peek out from between leafy floral bracts. Our species, although widespread, is often overlooked and may be endangered in some places because its habitat is threatened by development, plowing, and grazing.

Self-heal (*Prunella vulgaris*).

There is some question as to whether the self-heal we find in grassy fields is native here or is a European import. Renowned in times past as a healing herb that promotes overall good health, self-heal is seldom used anymore. The oval, dark green leaves are unusual in the mint family for lacking a minty smell, but the dense spikes of tiered flowers are clearly mintlike in appearance. Deep royal blue flowers peek out from between fringed, dark purple floral bracts.

Chia (*Salvia columbariae*). P. 306.

Chia is famed among early California settlers and California Native Americans; the nutritious seeds are still avidly sought as a source of high energy and were also soaked for an eye wash. In years of abundant winter rains, chia grows by the hundreds on rocky or grassy slopes; then flowering stalks will top two feet and carry several tiers of flowers. Each tier has long, spiny, dark purple bracts and pretty blue flowers, the lower petals with darker dots set in a white patch. Chia is a lover of warm, sunny climates and becomes truly abundant to the south of us, often in warm deserts.

Hummingbird Sage (*Salvia spathacea*). P. 306.

Here is an entirely different salvia. Hummingbird sage is a perennial with creeping rootstocks and eventually grows into large colonies, often in the dappled shade of oaks. The long, elliptical, pale green leaves are heavily quilted and carry a lovely fragrance. Flower spikes appear throughout the year but are most visible in late spring. They carry large whorls of showy, rose-red flowers with long tubes and abundant nectar, a favorite for hummingbirds. Rare with us but growing commonly in the south, they naturalize beautifully under oaks in gardens, where their flowers help keep hummingbirds fed.

Yerba Buena (*Satureja douglasii*). P. 306.

Yerba buena was the original name for San Francisco, and the name still applies to the island in the middle of the bay. This little herb was a favorite remedy among the early Spaniards and its name translates as "good herb." Steeped in hot water, the leaves make a refreshing mint tea. The genus name "Satureja" comprises several species; the culinary herb we call savory belongs here but has its homeland in the Mediterranean region of the Old World. Yerba buena is hard to spot, for the stems creep flat along the ground in partial shade, often just by the edge of shrubs but not under them. The pale green, rounded leaves are prettily fluted, and if you look closely, there are minute white blossoms in between leaf and stem.

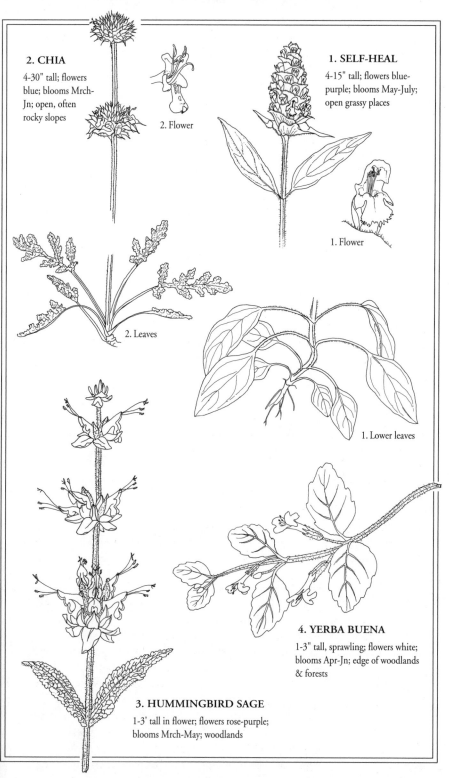

2. CHIA

4-30" tall; flowers blue; blooms Mrch-Jn; open, often rocky slopes

2. Flower

1. SELF-HEAL

4-15" tall; flowers blue-purple; blooms May-July; open grassy places

1. Flower

2. Leaves

1. Lower leaves

4. YERBA BUENA

1-3" tall, sprawling; flowers white; blooms Apr-Jn; edge of woodlands & forests

3. HUMMINGBIRD SAGE

1-3' tall in flower; flowers rose-purple; blooms Mrch-May; woodlands

Skullcap (*Scutellaria californica* and *tuberosa*). P. 307.

Skullcaps are so named because the sepals look like the old-fashioned skullcaps people used to wear to bed. The first impression of the whole flower, however, is of blue, purple, or white snapdragons, for the two lips come together in the same fashion as they do in the garden snapdragon. Yet skullcaps belong to the mint family, a fact not easily discerned from the unscented leaves. The safest way to identify this family is by the four-lobed ovary (two-chambered in snapdragons). Skullcaps are among our most charming woodland wildflowers, where they often brighten mossy embankments near ferns. *Scutellaria tuberosa* (tuberous skullcap) has tiny tubers in the soil and rich blue-purple flowers; *S. californica* (California skullcap) has nontuberous roots and white flowers.

Wood Mints (*Stachys* spp.). P. 307.

An alternate name for these plants is hedge nettle, owing to a superficial resemblance of the leaves to those of stinging nettle, but there is no close relationship, and this name is misleading. Wood mints do not have stinging hairs; instead the quilted leaves have an intense and strong odor unique to the genus; some find it alluring, others unpleasant. Wood mints all have widely wandering roots, so that in time large colonies are established. Our several species often look closely similar with their whorls of white, pink, or red-purple flowers. The lower lip has beautiful lines and dots as nectar guides. Two species stand out: *S. albens* (white-flowered wood mint) has wonderfully white wooly leaves and white flowers; *S. pycnantha* (rock wood mint) has the tiers of flowers spaced closely together. Both favor wet places along streams or seeps. Our most common wood mint is the dry-growing, pink-flowered *S. ajugoides rigida* (pictured).

Vinegar Weed (*Trichostema lanceolatum*). P. 307.

Despite its intense smell (reminiscent of turpentine or strong vinegar), this is a charming annual wildflower that blooms after soils have turned bone dry. Commonly bordering temporarily wet spots, vinegar weed doesn't really get going until the summer sun heats things up. The diminutive plants keep from being eaten by their intense odor; all you need is to be in the vicinity on a hot day to smell them. Up close the miniature blue blossoms have long, curled stamens that are reminiscent of their shrubby brother, wooly blue curls (*Trichostema lanatum*).

MUSTARD FAMILY (BRASSICACEAE).

Nonwoody plants with variable leaves (often with peppery taste). Flowers small and in racemes or spikes: 4 sepals, 4 separate petals narrowed at their bases, 6 stamens of 2 different lengths, single pistil with superior ovary, separated into 2 lengthwise partitions by a middle parchmentlike wall. Fruit a silique or silicle.

Rock Cress (*Arabis* spp.). P. 300.

Rock cresses are so named because they bear cresslike leaves (said to have peppery flavor) and they grow in rocky places where there's little competition. Look for arabises in places favored by rock ferns, saxifrages, penstemons, and dudleyas. Out of flower the low leaf rosettes are not particularly distinctive; in flower the modest purple, pink, or white blossoms are seldom showy but are pretty up close. The most interesting part of the plant is the long, slender seed pods, which may be straight, arched, pointed downwards, inclined upwards, or pressed against the stem. These different conformations are useful in identifying species. Our prettiest species is Brewer's rock cress (*Arabis breweri*), with dense leaf rosettes tucked against rocks on top of Mt. Diablo and pretty rose-purple flowers. Our most widespread arabis is tower mustard (*A. glabra*), with four- to six-foot stems bearing inconspicuous white flowers and clasping leaves.

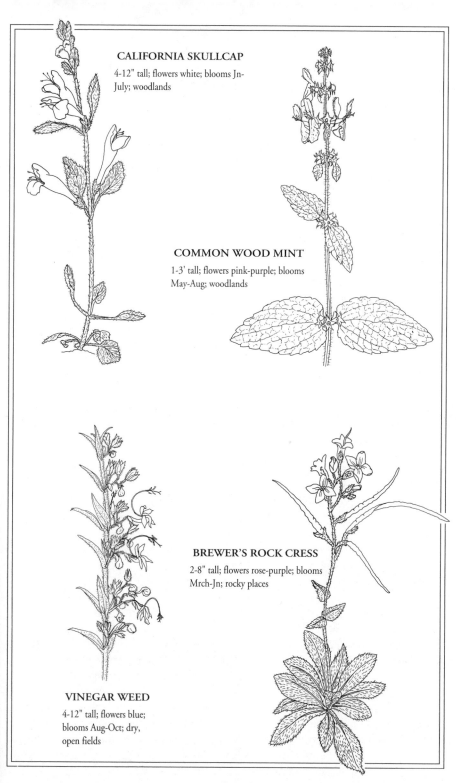

CALIFORNIA SKULLCAP

4-12" tall; flowers white; blooms Jn-July; woodlands

COMMON WOOD MINT

1-3' tall; flowers pink-purple; blooms May-Aug; woodlands

BREWER'S ROCK CRESS

2-8" tall; flowers rose-purple; blooms Mrch-Jn; rocky places

VINEGAR WEED

4-12" tall; flowers blue; blooms Aug-Oct; dry, open fields

Winter Cress (*Barbarea orthoceras*). P. 300.

Look for winter cress along stream courses in the spring. The pinnately divided leaves are peculiar in that the terminal leaflet is much larger than the side leaflets. The flowers are bright yellow and closely resemble those of the various mustards (*Brassica* spp.).

Mustard (*Brassica* spp. and *Hirschfeldia incana*).

Mustards color open grassy fields, roadsides, and orchards with a carpet of yellow in late winter and spring. So abundant are they that many consider them among our prettiest native wildflowers, only they're not native at all. Mustards were probably first brought from Europe for their valuable seeds: the seeds are pressed and the extracted oils are combined with vinegar, herbs, and seasonings to make commercial mustard spreads. Our several species flower at different times. Most conspicuous early in the year is field mustard (*B. rapa*); finishing up the year, with much taller, more open flowering stalks and smaller flowers, is jointed mustard (*H. incana*). Mustards often mingle with another nonnative—radish weed (*Raphanus* spp.)—but the latter has purple, pink, pale yellow, or white flowers.

Shepherd's Purse (*Capsella bursa-pastoris*).

Shepherd's purse is another nonnative, this time a modest annual found throughout grazed grasslands. In flower it is seldom noticed, for the minute white flowers scarcely measure an eighth of an inch across. The fruit, however, is distinctive, like an upside down triangle or old-fashioned shepherd's purse; hence the common and species names.

Bitter Cress (*Cardamine oligosperma*).

This tiny annual is most likely to show up in wet places in your garden, although it also occurs in lightly shaded natural areas. The leaves are prettily divided into pinnately arranged leaflets, each leaflet rounded. The tiny white flowers are seldom showier than those of shepherd's purse, but unlike the seed pods of the latter the slender bitter cress seed pods explode when ripe, scattering their seeds everywhere possible. If you want to get rid of bitter cress, pull the plants out before they have a chance to bloom.

Milkmaids or Toothwort (*Dentaria* [*Cardamine*] *californica*). P. 302.

Milkmaids is among our earliest wildflowers and, in years with abundant rain and mild temperatures, may even flower by the new year. Look for it on banks under woodlands and forests. The leaves and flowering stalks come from tiny, pearllike deep-seated tubers (said to taste like pepper); in years when they don't bloom, each tuber sends up a single broad, rounded leaf; in years when flowering is favored, a stem emerges with narrow, deeply slashed leaves and a raceme of lovely white flowers (often flushed purple on the outside).

Western or Foothill Wallflower (*Erysimum capitatum*). P. 302.

The spirelike racemes of burnt orange to yellow-orange flowers immediately identify foothill wallflower, in mid- to late spring. Favoring light shade, these beautiful wildflowers thrive on steep banks where they keep company with Chinese houses (*Collinsia heterophylla*) and clarkias. Although claimed to be perennial, they really are biennial, making a leaf rosette their first year and flowering and fruiting their second. The fragrant flowers are attractive to many pollinators. The name wallflower originated with the European kinds, for these commonly grew between rocks on monastery and castle walls.

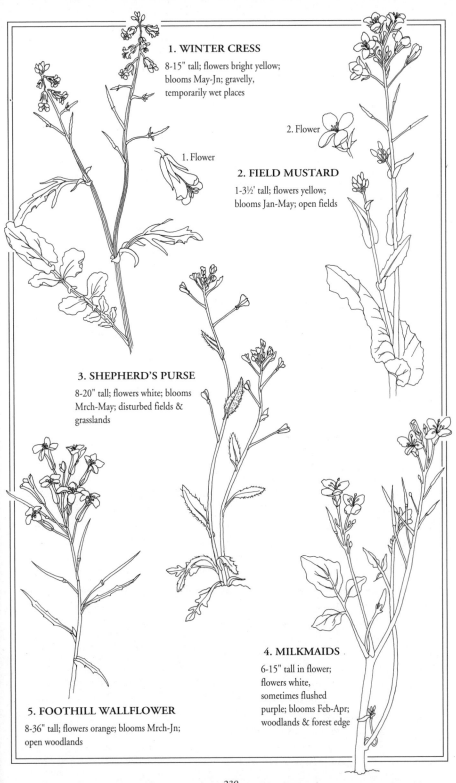

1. WINTER CRESS

8-15" tall; flowers bright yellow; blooms May-Jn; gravelly, temporarily wet places

1. Flower

2. Flower

2. FIELD MUSTARD

1-3½' tall; flowers yellow; blooms Jan-May; open fields

3. SHEPHERD'S PURSE

8-20" tall; flowers white; blooms Mrch-May; disturbed fields & grasslands

4. MILKMAIDS

6-15" tall in flower; flowers white, sometimes flushed purple; blooms Feb-Apr; woodlands & forest edge

5. FOOTHILL WALLFLOWER

8-36" tall; flowers orange; blooms Mrch-Jn; open woodlands

Pepper-grass (*Lepidium nitidum*). P. 303.

Pepper-grass is a third mustard relative with minute white flowers occurring in open grasslands. This species is native, as others are not, but the main identifying trait is once again the seed pod. These are circular, with a tiny notch in the top middle. Pepper-grasses often dry with the seed pods intact and make attractive dried arrangements. The foliage is said to taste peppery, as with so many other mustard relatives.

Water Cress (*Nasturtium officinale* [*Rorippa nasturtium-aquaticum*]).

Water cress is nearly ubiquitous in California's sluggish streams and marshes, even though it too originated in Europe. The plant you see growing naturally is the very same as the expensive green you buy in the supermarket, but don't collect water cress from the wild unless you're sure the water where it grows is uncontaminated. The deeply pinnately slashed leaves lie along the surface of the water, but the white blossoms are carried in showy racemes above the water. The genus name "nasturtium" is derived from two Latin words: "nas," nose, and "turt," to turn or twist, because of the strong flavor. This genus name is also the common name for a garden flower (whose peppery leaves and flowers are edible), but the latter is in the South American family Tropaeolaceae.

Radish Weed (*Raphanus raphanistrum* and *sativum*).

Radish weed colors large swaths of open habitat: disturbed pastures, grassy fields, and roadsides, where it often grows with wild mustard. Radish weed is not native; it is no more than the garden radish gone wild. When radishes interbreed for a while, they lose the succulent roots of the garden radish (roots become woody and inedible), yet the leaves and flowers look exactly like radishes that have stayed in the garden too long. The flowers come in many shades of purple, pink, and bronze or may be pale yellow or white, and the four petals beautifully exemplify the crosslike arrangement typical of the mustard family. Flowers and seed pods are edible and substitute for radish roots.

Jewel Flower (*Streptanthus* spp.). P. 307.

"Streptanthus" means twisted flower in Greek and alludes to the crimping and crisping of the narrow petals. The common name "jewel flower" is harder to interpret, but at close range the flowers often are beautifully colored. Much of the color actually comes from the sepals, which are deep wine-red, black-purple, or amethyst. Streptanthuses are not only unusual for their flowers, they grow in some of the most difficult habitats: loose rocky scree of volcanic origin, or blue serpentine. As such, the various species are highly restricted to these special rock "islands." The common jewel flower (*S. glandulosus*) occurs throughout the central Coast Ranges and has lavender-purple flowers; Mt. Diablo jewel flower (*S. hispidus*), on the other hand, is restricted to Mt. Diablo and is distinguished by its pale purple flowers and the abundance of stiff white hairs on leaves, seed pods, and stems. Look for it in burned-over areas and on loose scree.

Lacepod or Fringepod (*Thysanocarpus* spp.). P. 307.

Here is a fourth mustard that has nearly microscopic white flowers and is an annual in our grasslands. The seed pods are unquestionably the most beautiful of all, consisting of flattened discs with a single seed in the center. Sometimes red at maturity, the outer rim is crimped in fringepod (*T. curvipes*) and the fruits gracefully droop. In lacepod (*T. radians*) the seed pods are white with narrow spokes separated by tiny holes, and the fruits are held stiffly out from the stems. Probably the fruits on both species are carried frisbee-style by strong gusts when they're severed from the parent plant.

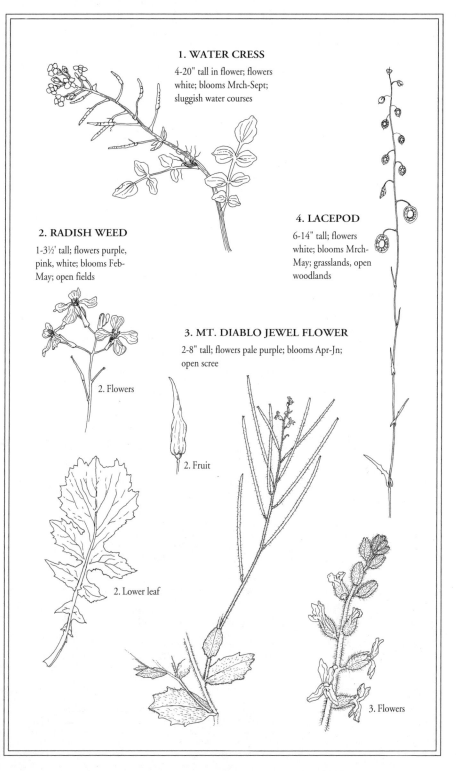

1. WATER CRESS

4-20" tall in flower; flowers white; blooms Mrch-Sept; sluggish water courses

2. RADISH WEED

1-3½' tall; flowers purple, pink, white; blooms Feb-May; open fields

2. Flowers

2. Lower leaf

4. LACEPOD

6-14" tall; flowers white; blooms Mrch-May; grasslands, open woodlands

3. MT. DIABLO JEWEL FLOWER

2-8" tall; flowers pale purple; blooms Apr-Jn; open scree

2. Fruit

3. Flowers

NETTLE FAMILY (URTICACEAE).

Nonwoody plants with undivided leaves, stipules, and often with stinging hairs. Flowers insignificant, tiny, greenish, in chains. Four sepals, no petals, 4 stamens, single pistil with superior ovary. Fruit a nutlet.

Stinging Nettle (*Urtica dioica*). P. 308.

Stinging nettle is more a plant to avoid than a wildflower to admire. It can be identified by its broad, coarsely saw-edged leaves with stipules; ribbed, hairy stems; and (at times) the remains of drooping flower chains. Other plants whose leaves superficially resemble nettle include California bee plant (but its leaves are not hairy and lack stipules) and wood mints (but their leaves are quilted and lack stipules). Strangely, the young leaves of stinging nettle make an excellent and nourishing cooked vegetable; gather them in early spring: wear gloves. The tiny green wind-pollinated flowers hardly merit a second glance.

NIGHTSHADE FAMILY (SOLANACEAE).

Small shrubs or nonwoody plants, leaves undivided and often foul smelling. Flowers trumpet to star shaped; 5 partly fused sepals, 5 much-fused petals, 5 stamens fused to petal tube and sometimes adhering to one another, single pistil with superior, 2-chambered ovary. Fruit a capsule or berry.

Jimson Weed or Angel's Trumpet (*Datura* spp.). P. 301.

Jimson weed has an unsavory reputation, for all parts of the plant are loaded with poisonous alkaloids except the flowers. Yet in mythology and legend these plants are renowned, for small amounts of these alkaloids have had long traditional use by native peoples for inducing trances, entering puberty rituals, and divination. These are stout plants with narrowly triangular, foul-smelling, dull green leaves and large, showy, trumpet-shaped white flowers, often flushed purple on the outside. The flowers open around late afternoon and waft a long-range perfume, for they're designed to attract evening-flying hawkmoths; they curl up the next morning. One flower is followed by another, then another. At season's end, flowers develop into curious, nodding, spine-covered seed pods (hence the additional common name "thorn apple"). Our two daturas are *D. stramonium*, an introduced species found in orchards and roadsides, and *D. wrightii* (pictured), doubtfully native here, favoring open canyon bottoms with sandy or gravelly soils. The latter is the showier of the two with flowers reaching six inches in length.

Wild Tobacco (*Nicotiana bigelovii* [*N. quadrivalvis*]). P. 305.

Our area has two very different wild tobaccos: *N. bigelovii*, a true native with clammily sticky leaves and stems and long slender tubular white flowers; and *N. glauca*, the tree tobacco, a Brazilian species that has entered gravelly ditches and dry arroyos (see p. 100). The genus name was given in honor of a Frenchman by the name of Nicot, and it also commemorates the poisonous substance nicotine, which is concentrated in leaves and stems. Native Americans used the dried leaves for smoking, but only for ceremonial purposes. The gardener may note a close resemblance between wild tobacco and the popular garden bedding plants called nicotiana.

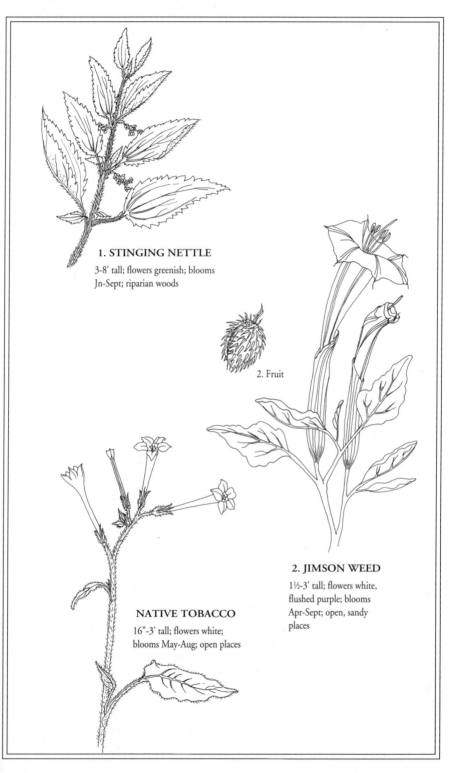

1. STINGING NETTLE

3-8' tall; flowers greenish; blooms
Jn-Sept; riparian woods

2. Fruit

2. JIMSON WEED

1½-3' tall; flowers white,
flushed purple; blooms
Apr-Sept; open, sandy
places

NATIVE TOBACCO

16"-3' tall; flowers white;
blooms May-Aug; open places

Black Nightshade (*Solanum nigrum*).

Black nightshade is a common weed favoring dump sites, roadsides, and gardens with rich soils. The plant branches a great deal, giving it a bushy aspect although it is not in the least woody. The small, white, starlike flowers are pretty and superficially look a bit like shooting stars, with their forward-pointing cone of yellow stamens and backward-curving petals. Stamens open at their tips and the pollen is vibrated out by the buzzing action of bees. The berries ripen to near-black; hence the common name. Some close relatives have been selected for cultivation under a variety of common names, such as bush huckleberry. Whether the berries are fully safe to eat is dubious, although we do eat the berries of the nonnative solanum called eggplant. The fruits are fully intended to attract birds, and they succeed too well in spreading the seeds far and wide.

ORCHID FAMILY (ORCHIDACEAE).

Nonwoody plants from bulbous or tuberous roots, leaves undivided and parallel veined. Flowers single, in long spikes or in racemes; irregular. Three colored sepals; 2 petals often resembling the sepals, and a third lower petal enlarged into a lip or pouch; 1 or 2 stamens fused to the style and stigma (which form the column); single pistil with inferior ovary, and thousands of minute seeds. Fruit a capsule.

Spotted Coral Root (*Corallorhiza maculata*). P. 301.

When it first pushes through the leaf litter in dimly lit forests, coral root resembles a fungus. The root looks like a piece of branched coral and is hooked up to saprophytic fungi that feed it. Without this fungal connection the plant would soon die, for as the shoot develops there are no green leaves, only reduced red-purple scales and flowers. The small flowers are miniature orchids, the sepals and two upper petals purplish. The lip is purple-spotted white and the reason for the species name, "maculata." Strangely, East Bay forms of this orchid lack these spots.

Brook Orchid (*Epipactis gigantea*). P. 302.

Brook orchid is widely distributed through the West, but it is not common in our area. You can consider yourself indeed fortunate to find a clump. It grows into large colonies with time, the rhizomes running between boulders of permanent brooks. There they are assured a constant water supply. The young shoots look much like those of false Solomon's seals, but by late spring or summer a raceme of curious and enchanting orchidlike flowers finally opens. The colors are unlike those of any other plant, with combinations of green, brown, maroon, and yellow, and the lip is hinged and movable (hence a second name: "chatterbox").

Rein Orchid (*Piperia* spp.). P. 306.

Rein orchids have such small flowers, you have to look closely to see their orchid design. In fact, with our species you're much more likely to notice the two to four tongue-shaped leaves that arise in early spring from tuberous roots. You must be patient to find the flowers, for just as the leaves are withering, a flowering spike shoots up. By early summer, there are dozens of tiny greenish to white flowers in a dense spike; each flower buries its nectar in a slender spur. In fact, it is the details of spurs as well as flower colors that distinguish the various piperias of the Coast Ranges.

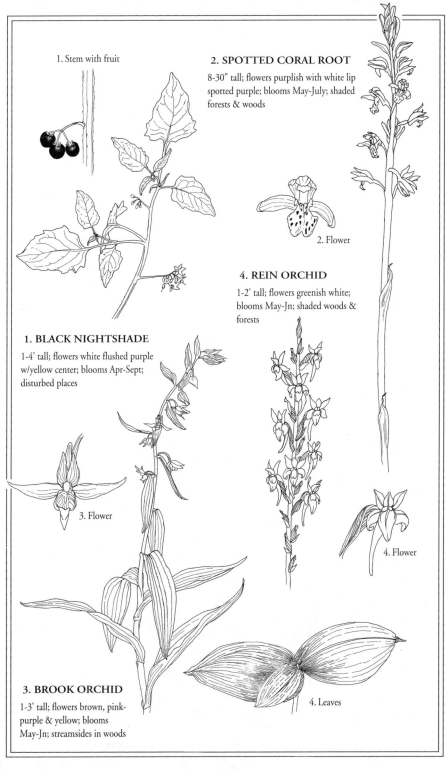

1. Stem with fruit

2. SPOTTED CORAL ROOT

8-30" tall; flowers purplish with white lip spotted purple; blooms May-July; shaded forests & woods

2. Flower

4. REIN ORCHID

1-2' tall; flowers greenish white; blooms May-Jn; shaded woods & forests

1. BLACK NIGHTSHADE

1-4' tall; flowers white flushed purple w/yellow center; blooms Apr-Sept; disturbed places

3. Flower

4. Flower

4. Leaves

3. BROOK ORCHID

1-3' tall; flowers brown, pink-purple & yellow; blooms May-Jn; streamsides in woods

PARSLEY FAMILY (APIACEAE).

Nonwoody, mostly perennial plants, often from thickened roots or tubers; the leaves sheathed at base, often lobed to deeply divided, usually fragrant when crushed (odor of anise, parsley, carrots, or other). Flowers small but borne in large numbers in compound umbrellas or umbels; 5 minute sepals, 5 separate petals, 5 stamens, single pistil with inferior ovary, and 2 styles and stigmas. Fruit separating into 2, 1-seeded parts.

Wooly Angelica (*Angelica tomentosa*). P. 299.

Angelica and cow parsnip are two members of the family that truly qualify as giant herbs, growing anew each year from long taproots and reaching several feet in height, with large compound leaves. Angelica leaves are softly wool covered, and have a smell unlike anything else except other angelicas. The European species was used medicinally, and the leaf stem was candied for decorating pastries. Like so many members of its family angelica has white flowers, and these are borne in umbels often a foot or more across.

Water Parsnip (*Berula erecta*). P. 300.

Water parsnip is another giant of the parsley family, also with large umbels of white flowers. You should note that many botanists identify these plants on the basis of their fruits, although fruits are usually missing at the time the plants are flowering. Water parsnip has barely ridged (ribbed) fruits, whereas angelica has easy-to-see ribs all around. The habitats are useful here, for angelica prefers dry, lightly shaded banks, whereas water parsnip lives in sluggish streams and marshes. It is uncommon.

Water Hemlock (*Cicuta douglasii*). P. 301.

It is nowhere more apparent than here that the parsley family has both edible and useful plants (such as caraway, anise, parsley, and carrots) and deadly poisonous ones. Water hemlock belongs to the latter category and is among the most virulently poisonous of plants. Even eating a small piece of the chambered root may bring convulsions and death! Unfortunately, water hemlock often grows with water cress, so picking any streamside plant for food should be done with great care. The tall stems bear coarsely twice-divided leaves and umbels of white flowers. (These umbels are considerably smaller than those of angelica, water parsnip, or cow parsnip.)

Poison Hemlock (*Conium maculatum*).

Although in the same family as water hemlock, poison hemlock belongs to a different genus and comes to us from Europe. How it was introduced is difficult to say, but it is one of our most abundant weeds along roadsides and in disturbed fields. Unlike water hemlock it prefers dry soils. Socrates was said to have taken this when he died, for poison hemlock is true to its name. Poison hemlock is an attractive plant with fernlike leaves and tall stems to six or more feet (but be aware that they may be much shorter in dry soils). The stems have purple blotches all over (hence the species name, "maculatum"), and there are small umbels of white flowers. The purple spots and the unpleasant odor serve to identify this plant.

Wild carrot (*Daucus pusillus*). P. 302.

Delicate and ephemeral, wild carrot will seldom be noticed in the field, even though it has relatively broad distribution in dry, open places on sandy or rocky soils. Our only native carrot relative, it is entirely scaled down from its showy sister Queen Anne's lace (*D. carota*). The latter is an abundant roadside weed in the north Bay and represents the garden carrot reverted

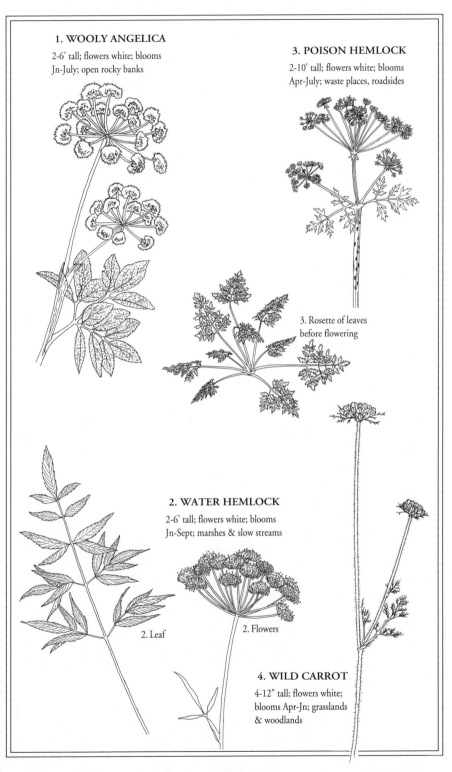

1. WOOLY ANGELICA

2-6' tall; flowers white; blooms
Jn-July; open rocky banks

3. POISON HEMLOCK

2-10' tall; flowers white; blooms
Apr-July; waste places, roadsides

3. Rosette of leaves
before flowering

2. WATER HEMLOCK

2-6' tall; flowers white; blooms
Jn-Sept; marshes & slow streams

2. Leaf

2. Flowers

4. WILD CARROT

4-12" tall; flowers white;
blooms Apr-Jn; grasslands
& woodlands

to its wild form. Queen Anne's lace, owing to its bold stature, is sometimes mistaken for poison hemlock but lacks the purple spots on its stems. Wild carrot, by contrast, stands no more than a foot tall, with widely diverging bristly-haired branches and modest umbels of tiny white flowers. Recognition is assured when you find the bristly-haired fruits or crush the foliage and smell carrot.

Fennel (*Foeniculum vulgare*).

Fennel is incorrectly called anise (true anise is related but grows only in cultivation). The Italians who brought this plant to the Bay Area are fond of a special form of fennel known as finnochio (also called fennel bulbs) that is cooked as a vegetable. The cultivated form reverted to its wild state when it escaped cultivation, and now it is one of our most abundant roadside weeds. Every part of the plant is scented like anise, from the feathery highly divided fernlike leaves to the tall stalks reaching over six feet and to the numerous umbels of yellow flowers. Plant it in your garden at your own risk; every seed will grow. The leaves and seeds are useful for flavoring.

Cow Parsnip (*Heracleum lanatum*). P. 303.

Cow parsnip is a ragged giant herb much resembling angelica. Its leaves, though, are coarser and only divided a few times, and its smell is reminiscent of strong parsnips. The giant hollow stems carry the largest umbels of white flowers in the family. The outer flowers are enlarged as compared with the inner, a sort of division of labor in which the outer flowers do the attracting of pollinators while the inner bear the nectar and pollen. Native Americans are said to have burned the leaves and used the ash as a salt substitute. Cow parsnip is only common near the coast or Bay, but it is occasional in protected canyons as far east as Mt. Diablo.

Biscuit Root (*Lomatium* spp.). P. 304.

Most lomatiums are low-growing perennial herbs from long starchy taproots, and most bear close tufts of finely divided, fernlike leaves. The scent is reminiscent of parsley, but the Native Americans were said to have used the roots, baking them for food. Flowering in mid-spring, most lomatiums have umbels of yellow flowers held only a few inches above the leaves. Perhaps our most widespread species is *L. utriculatum* and a close look-alike, *L. caruifolium* (pictured), common in grasslands, with bright yellow flowers. *Lomatium dasycarpum* occurs in similar circumstances but often on shallow, rocky soils, and it has pale yellow to creamy flowers. Finally, the distinctive *L. californicum* has coarsely divided bluish green leaves, lives on shaded banks, and grows taller, with whitish flowers and a celerylike odor. It was reputedly used by Native Americans for its medicinal properties.

Water Parsley (*Oenanthe sarmentosa*). P. 305.

Here is another parsley relative of sluggish streams and marshes, but unlike water hemlock, water parsley has creeping stems that root as they grow. The leaves are divided somewhat in the fashion of coarse Italian parsley, and the flowers are white. Perhaps most distinctive are the fruits, which often turn a deep wine-red; hence the genus name from Greek, "enos" (wine) and "anthos" (flower). Water parsley is reputed to be poisonous.

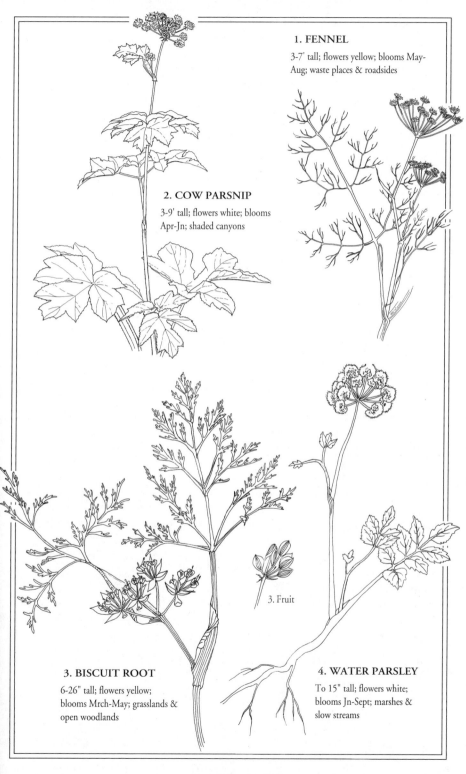

1. FENNEL

3-7' tall; flowers yellow; blooms May-Aug; waste places & roadsides

2. COW PARSNIP

3-9' tall; flowers white; blooms Apr-Jn; shaded canyons

3. Fruit

3. BISCUIT ROOT

6-26" tall; flowers yellow; blooms Mrch-May; grasslands & open woodlands

4. WATER PARSLEY

To 15" tall; flowers white; blooms Jn-Sept; marshes & slow streams

Sweet Cicely (*Osmorhiza chilensis*). P. 305.

Sweet cicely is a seldom identified but common woodland perennial herb, with modest, prettily divided leaves and small umbels of white flowers. It's most memorable for its anise scented fruits, lined with recurved prickles that catch on clothing and fur as an efficient means of dispersal. The root is also anise scented, and some claim that fruits and roots are useable as seasoning. An uncommon sister species—*O. brachypoda*—has larger fernlike leaves with lacy texture, especially in their natural habitat, where they create an underpinning to darkly shaded bay-oak woodlands. Look for it in Morgan Territory.

Yampah (*Perideridia* spp.). P. 305.

Yampahs are but one more parsley relative with umbels of white flowers. Most yampahs are distinguished this way: the leaves have narrow, linear divisions spaced far apart; the flowers appear late, mostly in summer; and the roots consist of clusters of strong fibers or tubers. The tuberous-rooted ones are most common in mountain meadows, and these were eagerly gathered by the Native Americans as a favorite vegetable, cooked in earth ovens. Since mountain yampahs occur by the thousands, they were an excellent source of food. One of our local species—*P. gairdneri*—is rare but has these tuberous roots. The other—*P. kelloggii*—is common in serpentine grasslands but has tough, inedible, fibrous roots.

Sanicles (*Sanicula* spp.). P. 306.

Sanicles are among our most abundant spring wildflowers, though few have truly showy flowers. Woodland sanicle (*S. crassicaulis*) has palmately lobed leaves with a finely fringe-toothed margin and pale yellow flowers; tuberous and poison sanicles (*S. tuberosa* and *bipinnata*) have compound fernlike cilantro scented leaves (but don't eat them!) and bright yellow flowers; and purple sanicle (*S. bipinnatifida*) has coarsely divided bluish green leaves and dark red-purple (occasionally pale yellow) flowers. All sanicles are grouped together according to the dense headlike clusters of flowers (rather than flowers in umbellets) and the prickly, recurved barbs on the fruits (making them effective at animal-based dispersal).

Shepherd's Needles (*Scandix pecten-veneris*).

Shepherd's needles is an annual of disturbed grasslands, introduced from Europe. Its modest umbels of white flowers are seldom noticed, but the fruits are highly distinctive, with a long needle or beak attached. This beak, the source of the common and species names, catches on clothing and animal fur and is responsible for aiding the plant in its seed dispersal.

PEA FAMILY (FABACEAE).

Nonwoody to woody plants, roots with nitrogen-fixing nodules, leaves usually pinnately compound or divided into 3s; stipules present. Flowers irregular and pea shaped, with 1 banner petal, 2 side wing petals, and 2 fused keel petals; 5 partly fused sepals, usually 10 stamens with 9 fused together by their filaments; single pistil with superior, 1-chambered ovary. Fruit a pea-pod-like legume.

Gambel's Locoweed (*Astragalus gambellianus*). P. 300.

Locoweeds are anathema to livestock, and many species are implicated in causing serious debilities through various types of poisoning. Some species are selenium accumulators while others contain alkaloids. The name "locoweed" alludes to the symptoms of such poisonings, after which livestock appear to go crazy or "loco." Despite their evil reputation locoweeds are a fascinating group, with over ninety kinds found in California alone. They vary from small

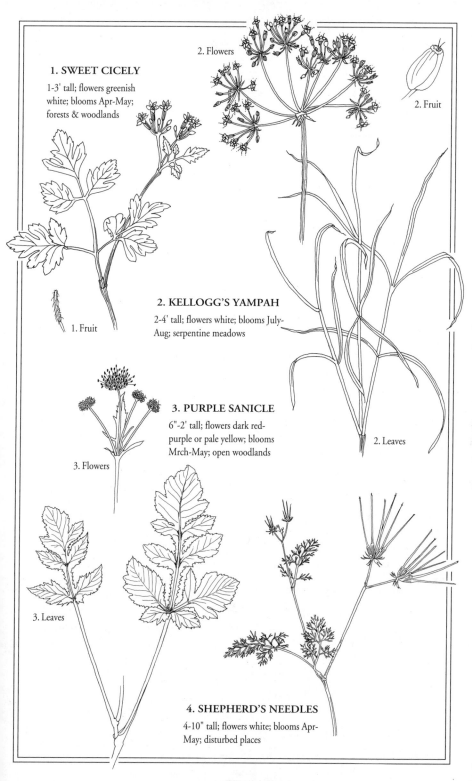

1. SWEET CICELY

1-3' tall; flowers greenish white; blooms Apr-May; forests & woodlands

2. Flowers

2. Fruit

1. Fruit

2. KELLOGG'S YAMPAH

2-4' tall; flowers white; blooms July-Aug; serpentine meadows

2. Leaves

3. PURPLE SANICLE

6"-2' tall; flowers dark red-purple or pale yellow; blooms Mrch-May; open woodlands

3. Flowers

3. Leaves

4. SHEPHERD'S NEEDLES

4-10" tall; flowers white; blooms Apr-May; disturbed places

annuals—such as Gambel's locoweed—to mat-forming alpines or bushy multistemmed perennials. Few are native to our area. In fact, all others listed here are considered rare. Most species favor deserts and other semidry habitats, such as rock scree in high mountains. Gambel's locoweed is an annual only inches tall, with typical pinnately compound pealike leaves and short, dense spikes of miniature pale purple pealike blossoms. Look for it tucked in among grasses in open fields.

Deer Broom Lotus or Deer Weed (*Lotus scoparius*). P. 304.

Deer broom lotus is technically a subshrub, with woody base, but the sprawling, broomlike green twigs hardly suggest woodiness. Deer broom is well adapted to hot dry summers along the margins of chaparral, and it germinates abundantly after fires. It manages to survive the summers by shedding the small, three-part leaves; only the branches carry on photosynthesis until rains return. Pretty umbels of yellow pea flowers adorn branches in spring, and these later fade to a striking red hue.

Lotuses or Bird's Foot Trefoil (*Lotus* spp.). P. 304.

Our other lotuses are not woody, instead varying from tiny, summer-blooming annuals to bushy perennials. Incidentally, the genus name "lotus" should not be confused with the common name "lotus" that is reserved for a genus (*Nelumbo*) related to waterlilies. Notable lotuses include *L. crassifolius*, a bushy robust perennial with odd purplish-green flowers, illustrated here; *L. corniculatus* (bird's foot trefoil), a ground-hugging short-lived perennial introduced to stabilize banks near the coast (or in the fog zone), with bright yellow flowers; *L. micranthus*, a low sprawling summer-blooming annual with single tiny, pale pink flowers; *L. subpinnatus* [*L. wrangelianus*] a mat-forming or sprawling annual with short hairs and yellow flowers; and *L. humistratus* (colchita), another matlike annual, with shaggy white hairs and yellow flowers fading red. All lotuses have asymmetrical trifoliate to pinnately compound leaves, with flowers borne in the angles between stem and leaf or in umbels.

Lupines (*Lupinus* spp.). P. 304.

Lupines are one of the most important native genera in the pea family, with dozens of species, several occurring in our area. (See the shrub section for bushy lupines.) The name "lupine" comes to us from the Latin for wolf, as the roots were believed to rob the soil of nutrients just as wolves robbed farmers of their chickens. Actually, they had it backwards; lupines often live in nutrient-poor soils (such as sands and rocky banks) because the nitrogen-fixing nodules on their roots have bacteria inside that convert soil nitrogen to a form usable by plants. Lupines are readily identified by their palmately compound leaves, where the leaflets are arrayed like fingers on a hand (other legumes are trifoliate or pinnately compound). Flowers are borne in showy spikelike racemes and are responsible for some of the best displays of spring wildflowers. Difficult to key to species, here are some common kinds for our area. *L. bicolor* (dove lupine) is a tiny annual with miniature blue and white flowers. *L. nanus* (sky lupine) is a taller annual with blue to blue-purple and white flowers of much greater size (and one of our most typical wildflowers, with California poppies for a blue-and-gold theme). *L. succulentus* (arroyo lupine) is a stout annual with hollow stems and showy spires of dark blue flowers. *L. formosus* (wood-land lupine) is a hairy perennial with purple flowers. *L. densiflorus* [*L. microcarpus densiflorus*] is a distinctive annual with precise tiers of flowers, the flowers colored white or cream and occasionally pink, lavender, or bright yellow. *L. subvexus* [*L. microcarpus microcarpus*] is similar to the last but with closely held whorls of red-purple flowers.

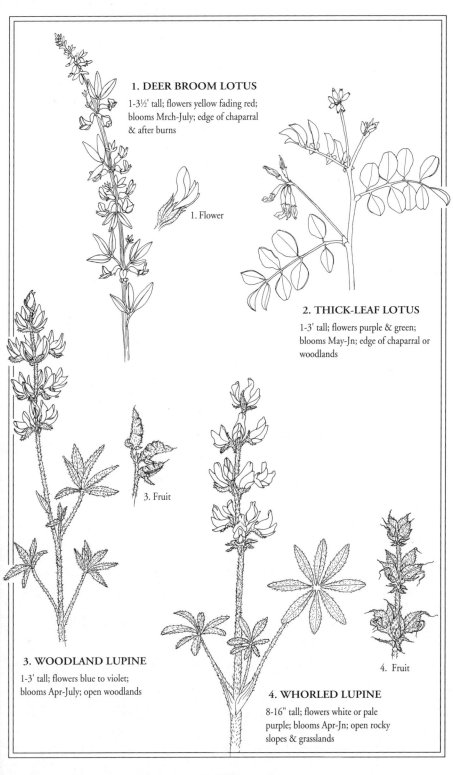

1. DEER BROOM LOTUS

1-3½' tall; flowers yellow fading red;
blooms Mrch-July; edge of chaparral
& after burns

1. Flower

2. THICK-LEAF LOTUS

1-3' tall; flowers purple & green;
blooms May-Jn; edge of chaparral or
woodlands

3. Fruit

3. WOODLAND LUPINE

1-3' tall; flowers blue to violet;
blooms Apr-July; open woodlands

4. Fruit

4. WHORLED LUPINE

8-16" tall; flowers white or pale
purple; blooms Apr-Jn; open rocky
slopes & grasslands

Lupines are one of our best wildflowers to naturalize in the garden from seeds, but be sure to soak seeds thoroughly before planting and guard against predation by slugs and snails.

Bur Clover and Alfalfa (*Medicago polymorpha* and *sativa*).

All medicagos come to us from the Mediterranean region. Some are accidental wayfarers that are now invasive weeds of disturbed grasslands and gardens (bur clover), others deliberately introduced for fodder and hay (such as alfalfa). All medicagos have leaflets in threes (similar to those of true clovers), but the fruits are unusual: instead of forming a straight pea pod, they are tightly coiled up. Those of the bur clovers are lined with sharp, recurved prickles that allow them to detach and ride on human clothing and animal fur, a real nuisance and reason for their great success. Bur clovers have tiny, bright yellow pea flowers before they make these burs; one distinctive kind—*M. arabica*—has a black spot on each leaflet. Alfalfa, on the other hand, has unspiny pods and deep blue-purple flowers in short spikes. Alfalfa is one of the best crops for fodder and is rich in protein, but the plants have thirsty roots, and their cultivation in the West is not appropriate for that reason. You'll see alfalfa naturalized along roadsides.

Sweet Clover (*Melilotus alba* and *indica*).

The sweet clovers have, like alfalfa, found use as protein-rich sources of food for livestock, and their sweetly scented flowers are bee favorites widely used by honey husbandmen. Like the medicagos they're strictly from the Old World, but they are now so widely naturalized they seem native. The two common species are white sweet clover (*M. alba*) and yellow sweet clover (*M. indica*). Both have narrow spirelike clusters of small pea flowers, whereas the true clovers have flowers arranged in heads.

California Tea and Indian Hemp (*Psoralea physodes* [*Rupertia physodes*] and *P. macrostachya* [*Hoita macrostachya*]). P. 306.

The psoraleas are a seldom noticed group of legumes with interesting appearance and use. Most have bold leaves divided into three ovate to nearly round leaflets and foliage dotted with dark oil glands, releasing a strong resinous scent when crushed. The flowers are borne in short to long spikes. California tea is a low, bushy plant with strong odor, pointed leaflets, and curious greenish white flowers. It is scattered in forests but nowhere typical. The common name probably alludes to early use of the leaves for tea, but if so, the tea must have tasted strongly medicinal. Our second species—Indian hemp—is a stream follower with tall, straight stalks to over eight feet, less highly scented leaves, and spikes of pale purple flowers. The strong fibers of the stems were used for rope and twine by California Native Americans, much in the manner of the unrelated Indian hemp, *Apocynum cannabinum*. A third uncommon species—*P. californica* [*Pediomelum californica*]—found atop Mt. Diablo, looks distinctly like a lupine, for the leaves are palmately compound. It is easily distinguished at close range by the almost-headlike clusters of pale blue-purple and white flowers and the distinctive odor produced by oil glands.

Clovers (*Trifolium* spp.). P. 307.

The true clovers are a large, important group, mostly of annuals, with leaflets in threes (think of shamrock) and heads of tiny but numerous densely packed flowers. Flower colors range from pure white through pinks and purples with an occasional yellow. The clovers you see in cultivated fields, disturbed and grazed grasslands, or along roadsides are usually European in origin, and were brought in as another good protein source for hay and fodder. Three commonly seen nonnatives are rose clover (*T. hirtum*), with large heads of deep rose-purple

1. BUR CLOVER

3-6" tall, plant sprawls;
flowers yellow; blooms
Mrch-Jn; waste places

1. Fruit

2. Flower

2. WHITE SWEET CLOVER

3-6' tall; flowers white; blooms May-
Aug; disturbed fields

3. CALIFORNIA PSORALEA

3-6' tall; flowers purple; blooms May-Jn;
stony places on edge of chaparral

4. TOMCAT CLOVER

4-16" tall; flowers pale pink-purple
& white; blooms Mrch-May;
grasslands & open woodlands

flowers (a bee favorite); white clover (*T. repens*), a creeping clover with white flowers common in lawns; and hop clover (*T. dubium*), a slender annual with small thimblelike clusters of yellow flowers. We have several native species as well, many of which were considered a seasonal food treat by Native Americans. Fondness for clover blossoms was said to be such a passion that sometimes bloating would result. Native clovers are most frequently found in grasslands and swales. Tomcat clover (*T. tridentatum* [*T. willdenovii*]) has slender leaflets and modest heads of purple and rose flowers. Small-head clover (*T. microcephalum*) has broad heads of pink flowers and fuzzy leaves. Seep clover (*T. wormskioldii*) is a creeping perennial with pretty heads of red-purple flowers. Fragrant clover (*T. variegatum*) is an annual abundant in vernally wet spots and has deliciously sweet-smelling heads of pink-purple flowers tipped white. Cow clover (*T. fucatum*) is a coarse plant of wet ditches, with pale yellow and reddish flowers that inflate in fruit. Last, balloon clover (*T. depauperatum*) is a tiny annual, with especially small heads of red-purple flowers that also inflate in fruit.

PHLOX FAMILY (POLEMONIACEAE).

Nonwoody plants or small shrubs, leaves variable. Flowers often showy; 5 partly fused sepals, 5 partly fused petals forming a short to long tube, 5 stamens fused to this tube, single pistil with superior ovary and 3-lobed stigma. Fruit a 3-chambered capsule.

Gilia (*Gilia capitata* and *tricolor*). P. 303.

Whether you pronounce it hee-lee-ah, ghil-lee-ah, or jill-ee-ah, the gilias are a diverse group of annuals preferring grasslands and deserts. Formerly the genus included other plants, which are now in separate genera: presently, gilias are characterized by their simple or pinnately slashed, alternate leaves. Our two most common gilias are quite different from each other. Bird's eye gilia (*G. tricolor*) is a slender plant only inches tall with several colorful but separated flowers. Each individual flower is a work of art: pale bluish petal lobes are outlined in darker purple, the entrance to the nectar tube having a deep purple ring and, within, a yellow bull's-eye. This color combination is a winner with bees. By contrast globe gilia (*G. capitata*) is a taller annual with dense, ball-shaped heads of powder blue flowers, occasionally ranging into darker shades or pure white. The entire flower is a uniform color. Despite the color differences, globe gilia is equally successful in attracting pollinators.

Linanthus (*Linanthus* spp.). P. 304.

The linanthuses are another group of annuals once placed in the same genus with gilias. A quick glance, however, will show the difference: linanthus leaves are in pairs, and each leaf is divided into several slender, fingerlike lobes. The generic name means linear ("lin") and flower ("anthus"), for the uncommonly long, linear flower tube that holds the nectar. Linanthuses are among our most abundant annuals, often of quite short stature but making up for it by sheer large numbers. Our most common species—*L. bicolor* (pictured here)—has heads of tiny starlike flowers in pure white or pale pink to lilac with a yellow throat. Similar in stature is the rare yellow-flowered *L. acicularis*. Other often-stouter linanthuses are *L. androsaceus*, a larger-flowered and taller species with a distinct golden throat; *L. grandiflorus*, a larger-flowered species with uniformly white or pink flowers; *L. ciliatus*, a larger-flowered white or pink species that has bracts and sepals margined with coarse white hairs; and the lovely *L. dichotomus* (evening snow), a low-growing species with flowers up to an inch across, fragrant at night and opening from late afternoon to evening. The white color and fragrance of the last attest to its attractiveness to moths.

1. COW CLOVER

3-6" tall; flowers pale
yellow tinted pink;
blooms Apr-May;
temporarily wet spots in
grasslands

2. GLOBE GILIA

8-32" tall; flowers pale blue;
blooms May-Jn; rocky slopes
& grasslands

3. Flower

3. BIRD'S EYE GILIA

4-16" tall; flowers pale blue
purple w/ dark purple center &
yellow throat; blooms Mrch-
Apr; grasslands

4. LINANTHUS

2-12" tall; flowers lilac,
pink, or white w/yellow
throat; blooms Apr-
May; grasslands &
open woodlands

Skunk Weed (*Navarretia* spp.). P. 305.

Many of the navarretias first call attention to themselves when we step on them and wonder where the skunk is. So closely do the leaves of this summer annual match the odor of skunks that it's difficult to be persuaded the odor is really from a plant. The odor, of course, has a wily protective function: no one wants to mess with a skunk! Most navarretias thrive in hard-packed soils and bloom at the end of the wet season, or they grow along the edges of vernal pools, where they burst into flower as soils dry out. Despite the strong odor of some species, some are also scentless. Other characteristics of navarretias are the long, spiny floral bracts around the heads of flowers and the spine-tipped sepals. The flowers themselves are quite small and clear blue, purple, or white, with narrow tubes.

Annual Phlox (*Phlox gracilis*). P. 306.

As showy as the masses of gilias and linanthuses are in spring, annual phlox is a demure and shy annual only inches tall, seldom in great numbers, and with only a few flowers open at any one time. The slender stems bear narrow, simple leaves and are topped by perfect miniature phlox flowers: five tiny pink petals with a minute notch at the end of each one. Botanists have long disagreed about where annual phlox belongs, for most true phloxes are perennials, often woody at the base and with larger, showier flowers. Some have moved it into the separate genus *Microsteris*, but the newest treatment considers it once again a true phlox. Look for annual phlox in openings of woods.

PINK FAMILY (CARYOPHYLLACEAE).

Nonwoody plants with opposite leaves and swollen nodes (feel the stem where the leaves are attached); papery stipules sometimes present. Flowers often borne singly; 5 separate or partly fused sepals, 5 separate petals narrowed at their base, 5 or 10 stamens, single pistil with superior ovary and variable number of styles. Fruit a single-chambered capsule with central stalk bearing seeds.

Douglas's Sandwort (*Arenaria* [*Minuartia*] *douglasii*). P. 300.

Sandwort—meaning a plant growing in sandy places—is most often thought of as a cushion-forming mountain plant, but Douglas's sandwort is an ephemeral, inches-tall annual with delicate, threadlike stems. Even the leaves are scaled down. Only the white flowers are relatively large for the size of the plant; after all, this is where the energy needs to go to attract pollinators. Look for this sandwort on shallow, rocky soils, where there's little competition. The threadlike stems in combination with the graceful white blossoms create an ethereal look. Another species—*Arenaria macrophylla*—is a rare, mat-forming perennial with modest white flowers. Look for it at the summit of Mt. Diablo.

Windmill Pinks (*Silene gallica*).

This nonnative little pink is actually a weed, but seldom a serious pest. Out of flower, the short stature and dull green, narrowly spoon-shaped leaves hardly get a second glance, but in flower and up close the plant is charming. The sepals are fused and striped with dark ribs, and the white to pale pink petals are twisted in the same fashion as the paper windmills children used to make (hence the common name). Most other silenes or pinks are perennial natives, as are those below.

4. WINDMILL PINKS

4-16" tall; flowers white to pale pink; blooms Mrch-Jn; open rocky spots or in grassland

4. Flower

2. ANNUAL PHLOX

3-8" tall; flowers pinkish; blooms Mrch-May; open woodlands

1. Flowers

1. SKUNK WEED

2-8" tall; flowers pale blue; blooms May-July; barren spots in grasslands

3. DOUGLAS'S SANDWORT

2-8" tall; flowers white; blooms Apr-Jn; rocky places, often serpentine

Indian Pink (*Silene californica*). P. 307.

The only other obvious pink in our area is the beautiful Indian pink. Dormant in the winter, the floppy stems with paired, pale green, fuzzy leaves in spring hardly suggest anything special to come, but by spring's end (and continuing into summer) a succession of oversized scarlet-red flowers open one by one. Each petal is deeply slashed, as are those of so many other, cultivated pinks; in fact, the flower of Indian pink rather resembles a fire-engine-red carnation. The brilliant color, slightly swept-back petals, long sepals (which protect the well of nectar), and protruding stamens all proclaim this a hummingbird flower. Indian pink continues to provide flashes of red after red larkspur has dried and gone to seed.

Spurreys (*Spergula arvensis*, and *Spergularia rubra* and *macrotheca*). P. 307.

Our two spurrey genera share many similarities: they're small plants with slender leaves and tiny flowers, the petals scarcely longer than the sepals. The true spurrey—*Spergula arvensis*—is a wayward plant from Europe, often abundant in sandy fields and orchards. It is easily identified by the whorled, nearly linear leaves and white flowers. Sand spurreys—*Spergularia rubra* and *macrotheca* (pictured)—form matted, sprawling plants, with small semifleshy leaves in small clusters and pale pink to lavender flowers. They also come from Europe, and they favor hard-packed soils along roadsides, by trails, or on the edge of disturbed fields (since in competition with taller plants they're not likely to survive). Other sand spurreys are native; a couple favor coastal bluffs or salty places near the ocean and have especially succulent leaves.

PLANTAIN FAMILY (PLANTAGINACEAE).

Nonwoody plants, mostly with basal leaf rosettes, leaves sometimes parallel veined. Flowers brownish to greenish, tiny, in spikes; 5 minute sepals, 5 papery petals, 5 stamens fused to petal tube, single pistil with superior ovary. Fruit a capsule.

Annual Plantain (*Plantago erecta*). P. 306.

The common plantains are pesky lawn weeds with taproots and nearly flat rosettes of leaves only too familiar to gardeners. These include English plantain—*P. lanceolata*—and broadleaf plantain—*P. major*—and come to us from Europe. But California has a very different-looking tiny annual species, abundant in grasslands in mid-spring. The first sign it's a plantain is when the tiny leaf rosettes appear, for the narrow fuzzy leaves have a grasslike appearance seen in the weedy kinds as well. More amazing is the short spike of whitish-green flowers, which is inconspicuous until you squat down and is lovely under a hand lens. Each flower has four slightly curled back parchment-paper-like petals and dark stamens. Whether it is insect pollinated is a mystery; the wind-pollinated weedy plantains have long protruding stamens to expose their pollen efficiently to winds; our native annual does not follow suit.

POPPY FAMILY (PAPAVERACEAE).

Nonwoody plants or small shrubs with acrid sap and variable leaves. Flowers usually large and showy, often single; 2 or 3 sepals that fall as flower opens, 4 or 6 petals (twice the number of sepals; often crumpled or textured), numerous stamens, single pistil with nearly superior ovary, variable stigmas. Fruit a capsule.

Prickly Poppy (*Argemone munita*). P. 300.

Prickly poppy lives up to its name, looking as though someone crossed a poppy with a thistle. All parts are spiny. Rare with us, prickly poppy appears occasionally on gravelly, disturbed

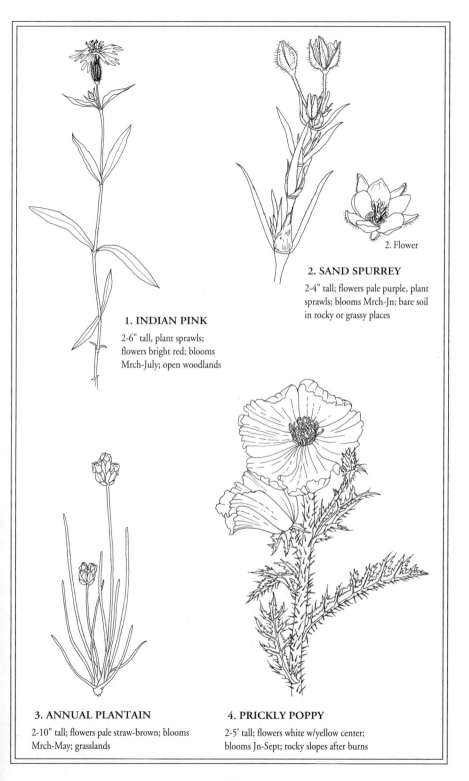

1. INDIAN PINK

2-6" tall, plant sprawls;
flowers bright red; blooms
Mrch-July; open woodlands

2. SAND SPURREY

2-4" tall; flowers pale purple, plant
sprawls; blooms Mrch-Jn; bare soil
in rocky or grassy places

2. Flower

3. ANNUAL PLANTAIN

2-10" tall; flowers pale straw-brown; blooms
Mrch-May; grasslands

4. PRICKLY POPPY

2-5' tall; flowers white w/yellow center;
blooms Jn-Sept; rocky slopes after burns

slopes, where it gains a competitive edge. The enormous white petals contrasted with the myriad yellow stamens make this one of our most striking, large-flowered natives, the overall effect similar to that of the flowers of the exotic Matilija poppy from southern California. One could justifiably apply another common name—fried egg plant—that is often applied to the Matilija poppy, alluding to the white and yellow color scheme of the round, flattened flowers.

California Poppy (*Eschscholzia californica*). P. 301.

Our state flower is one of California's most beautiful and characteristic wildflowers. In good years, grasslands wear a solid golden mantle of poppy flowers. This poppy has the curious habit of changing flower size with the season; flowers dwindle in brilliance and size as days lengthen, possibly a signal to changing pollinators. Like many of its sisters, California poppy contains opiates, and the root was said to be used by the California Native Americans for numbing toothaches. California poppy is one of the easiest wildflowers to grow in gardens, and it is available as seed from local nurseries. Plant during rains, and your poppies will flourish and resow themselves thereafter. Their efficiency at this process is due partly to the explosive seed pods.

Flame Poppy (*Papaver californicum*). P. 301.

Flame poppy is a very special wildflower, seldom seen except after fires. Unlike wind poppy, which blooms sporadically in the absence of fires, flame poppy seeds await the heat of fire to effect efficient germination, lying dormant in the ground for up to twenty years. Although the flowers superficially resemble those of wind poppy, the petals have green spots (not dark purple) at their bases, and the pistil has no style; instead the multiple stigma lobes fan out directly on top of the ovary or future seed pod.

Cream Cups (*Platystemon californicus*). P. 306.

Cream cups is one of our most lovely ephemeral annual wildflowers, with nodding buds—flushed pinkish—that by degrees right themselves as flowers open. Favoring sandy soils, some years they carpet the land with creamy saucers; other years they are scarce and scattered. On cloudy days petals curl together to protect stamens from getting wetted by rains; water destroys pollen in most flowers. Cream cups thrives on lack of competition; when annual grasses grow thick and tall, cream cups is stunted for lack of light.

Wind Poppy (*Stylomecon heterophylla*). P. 307.

The fragile-looking wind poppy is seldom common and, out of flower, the plants are easily overlooked. Unlike its cousin the California poppy, wind poppy seeks secluded places on steep slopes near dense woods or shrubs. Look for it to spring up in large numbers after burns; the hard seed coat is cracked by the heat of fire, allowing the imprisoned embryo plant to sprout the following spring when the rains return. This is one of our most beautiful poppies, but as its special requirements are difficult, it is not easy to maintain in the garden. Enjoy it as a special event in its natural haunts.

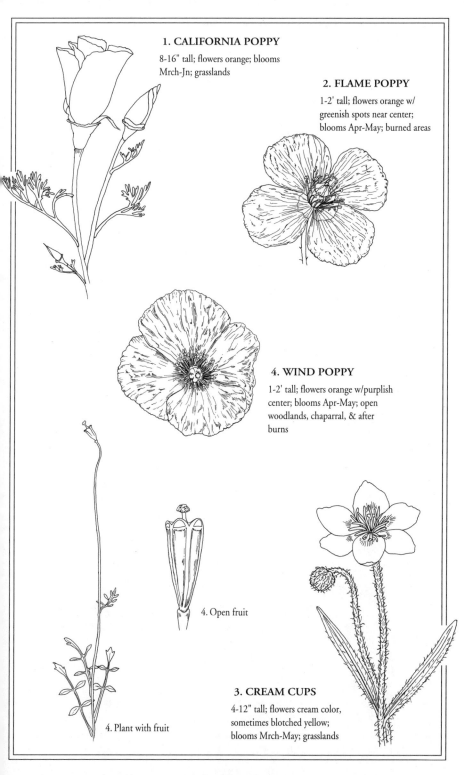

1. CALIFORNIA POPPY

8-16" tall; flowers orange; blooms
Mrch-Jn; grasslands

2. FLAME POPPY

1-2' tall; flowers orange w/
greenish spots near center;
blooms Apr-May; burned areas

4. WIND POPPY

1-2' tall; flowers orange w/purplish
center; blooms Apr-May; open
woodlands, chaparral, & after
burns

4. Open fruit

3. CREAM CUPS

4-12" tall; flowers cream color,
sometimes blotched yellow;
blooms Mrch-May; grasslands

4. Plant with fruit

PRIMROSE FAMILY (PRIMULACEAE).

Nonwoody plants with simple whorled, basal, or opposite leaves. Flowers often in umbels; 4 or 5 partly fused sepals, 4 or 5 slightly fused to clearly fused petals, 4 or 5 stamens fused at base to petals, single pistil with superior, single-chambered ovary. Fruit a capsule, often opening at the top.

Scarlet Pimpernel (*Anagallis arvensis*).

Everyone exclaims in pleasure when they first see the flowers of this pretty little European weed, for the flowers are indeed charming: miniature orange-red saucers eyed with deep purple and centered with short yellow stamens. The occasional blue-flowered variation is a pleasant counterpart and less aggressive. But if you have to live with this wildflower in your garden you soon see it as a prolific weed, since the near-prostrate stems stretch and turn ungainly, and the seed capsules keep popping off ever more seeds.

Shooting Stars (*Dodecatheon* spp.). P. 302.

You can hardly fail to be captivated by this lovely early spring perennial. Behaving like our bulbs, shooting star is dormant all summer and fall, sending up a low mat of oval, pale green leaves after winter rains, then naked flowering stalks with umbels of lovely pink or lavender flowers as spring days lengthen. The flowers indeed look like shooting stars, the petals swept abruptly backwards (and hiding the sepals), the near end capped by the fused, sharply pointed stamens. The base of the petals is also prettily detailed, with bands of white, yellow, and dark purple. When you see the larger-flowered garden cyclamens, you can't help notice the similarity in flower design; cyclamens also belong to the the primrose family but are native to the Mediterranean region. Our two shooting star species are *D. hendersonii* (Henderson's shooting stars), with inch-long, slender flowers; and *D. clevelandii* subspecies *patulum* (padre shooting stars), with somewhat larger, chunkier flowers. The latter species most often appears here in its white-flowered subspecific form, named above.

Star Flower (*Trientalis latifolia*). P. 307.

This third member of the primrose family appears as different from the other two wildflowers as possible. A lover of damp, dark forests, it grows from a small, deep-seated tuber and dies back to the ground from fall through winter, renewing its leaves each spring. The slender stem bears a whorl of from four to seven broad leaves at its tip, then follows these with two to five exquisite pale pink to almost white starlike flowers on threadlike stalks. Unlike others of its family, the petal number is not constant nor does it follow the usual four or five; normally, star flower has six, seven, or eight petals per flower. Look for it on the edges of redwood forests.

PURSLANE OR PORTULACA FAMILY (PORTULACACEAE).

Nonwoody plants with fleshy, succulent, simple leaves. Usually 2 separate sepals and 5 separate petals, 5 or 10 stamens, single pistil with variable number of styles, and superior ovary. Fruit a capsule; seeds often shiny and black.

Red Maids (*Calandrinia ciliata*). P. 300.

The vivid red-purple sheets of red maids in spring grasslands is unmistakable; this luminous color often contrasts splendidly with the cream of cream cups, orange of poppies, gold of goldfields, or blues of lupines. Up close, the clue that this is a bee flower despite the red overtones comes from the darker stripes on the petals, which serve as nectar guides, and the rich, yellow stamens; undoubtedly bee eyes see only the purple component of the floral

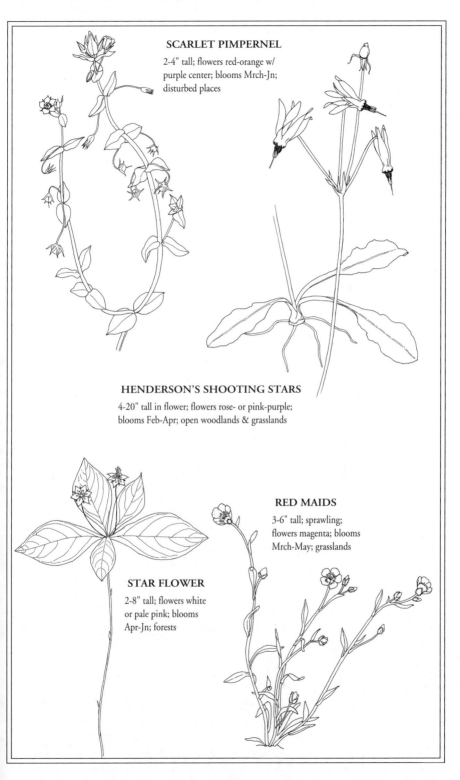

SCARLET PIMPERNEL

2-4" tall; flowers red-orange w/
purple center; blooms Mrch-Jn;
disturbed places

HENDERSON'S SHOOTING STARS

4-20" tall in flower; flowers rose- or pink-purple;
blooms Feb-Apr; open woodlands & grasslands

RED MAIDS

3-6" tall; sprawling;
flowers magenta; blooms
Mrch-May; grasslands

STAR FLOWER

2-8" tall; flowers white
or pale pink; blooms
Apr-Jn; forests

pigments rather than the red so apparent to us. Red maids seeds were favorite components of pinole, a mixture of seeds gathered seasonally by Native Americans.

Bitterroot (*Lewisia rediviva*). P. 304.

As rare as it is beautiful, bitterroot is worth the effort of looking for at the right time of year and in the right habitat. Because of its deep, carrotlike taproot, bitterroot is able to grow through tiny rock crevices on hot, inhospitable serpentine or lava rock outcroppings. Dormant half the year, the first sign of growth in spring is the curious rosette of narrow, almost tubular dull green leaves, but the plant is still difficult to see against its drab background until its two- to three-inch blossoms open. These appear during the first half of the day, each blossom an exquisite satiny rose, pink, or occasionally white saucer with overlapping petals and several cactuslike stamens. The beauty of the flowers soon fades, as sepals curl around the nascent seed pod, and within a couple of weeks the papery seed pods spread their seeds, and the plants retire for the remainder of the season.

Miner's Lettuce and Relatives (*Montia perfoliata* [*Claytonia perfoliata*] and spp.). P. 305.

Miner's lettuce has had long use as a leaf vegetable, for the crisp, succulent leaves have much the flavor of lettuce. Like lettuce they become bitter with age. Identified normally in its flowering stage by the pair of fused, disc-shaped leaves around the stem ("perfoliate" leaves), miner's lettuce starts life in early spring with a rosette of near-linear leaves. As new leaves are gradually added to the rosette they soon become spatula shaped. It's only when the central flowering stalk appears that the dislike leaf pair is produced. Just above, a slender raceme of small white flowers emerges. In full sun, miner's lettuce leaves turn bright red, and the taste then is sugary, for the red is a sign of excess sugars, some of which are converted to these protective pigments. Other montias in our area include *M. spathulata* [*Claytonia exigua*], similar to miner's lettuce but with the pair of stem leaves only partly joined, and *M. gypsophiloides* [*Claytonia g.*], called spring beauty, a delicate annual with grayish-green basal leaves, inconspicuous stem leaves, and exquisite fragrant, pink-striped flowers with notched petals. Look for the latter on loose rock scree—often serpentine.

Purslane (*Portulaca oleracea*).

Purslane is one wildflower more often encountered in manmade environments than in the "wild," for it's a common weed of farmyards and gardens everywhere. The sprawling red stems bear notably fleshy leaves, sometimes tinged red, which may be cooked as a vitamin-rich vegetable. These stems carry a long succession of pretty pale yellow flowers throughout the life of the plant.

ROCK-ROSE FAMILY (CISTACEAE).

Small shrubs, often woody only at the base, leaves opposite, simple, entire. Flowers borne singly; 5 separate sepals and petals, numerous stamens, and single pistil with superior ovary. Fruit a capsule.

Rush- or Sun-rose (*Helianthemum scoparium*). P. 303.

Common to the Mediterranean region, rock-roses and sun-roses are seen with increasing frequency in California gardens because of their thrifty use of water and prolific flowering. Few are acquainted with our own diminutive but pretty native rush-rose; technically it's a small shrub, but owing to size it's treated here as a woody-based perennial. Look for rush-rose in

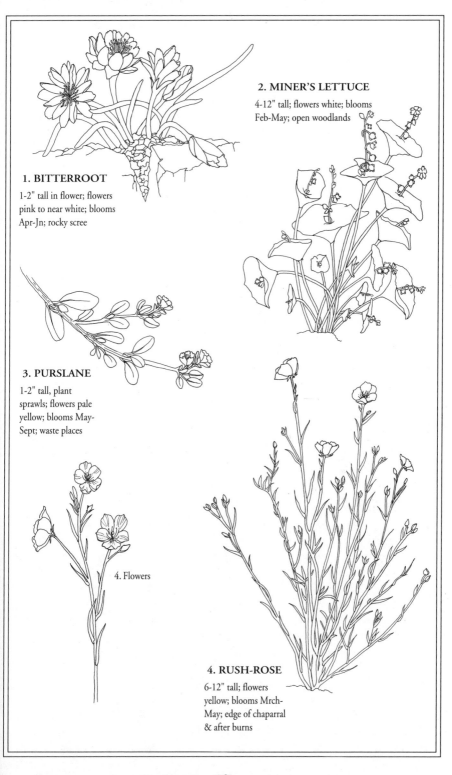

2. MINER'S LETTUCE

4-12" tall; flowers white; blooms
Feb-May; open woodlands

1. BITTERROOT

1-2" tall in flower; flowers
pink to near white; blooms
Apr-Jn; rocky scree

3. PURSLANE

1-2" tall, plant
sprawls; flowers pale
yellow; blooms May-
Sept; waste places

4. Flowers

4. RUSH-ROSE

6-12" tall; flowers
yellow; blooms Mrch-
May; edge of chaparral
& after burns

disturbed areas of chaparral, especially after burns, when it may appear by the hundreds; otherwise it is uncommon. The name rush-rose alludes to the brushy, broomlike green twigs and lack of obvious leaves (leaves are linear) and the pert, yellow blossoms looking like tiny roses.

ROSE FAMILY (ROSACEAE).

Nonwoody plants, shrubs, or small trees; leaves very variable, often with stipules. Flowers with 5 sepals (sometimes with an extra row of floral bracts resembling sepals) that form a cup, 5 separate petals, numerous stamens, and 1 to several pistils with ovaries superior, half-inferior, or fully inferior. Fruits fleshy (several types) or dry 1-seeded achenes.

Woodland Strawberry (*Fragaria vesca*). P. 303.

The woodland strawberry is a close counterpart to the famed French fraise des bois and is equally small-fruited, with concentrated taste. Like all strawberries, its leaves are borne in threes in low rosettes, and the long, trailing runners start new plants at their ends. Runners quickly spread woodland strawberry so that it establishes itself in minimal time as a low-maintenance, shade-tolerant ground cover for woodland gardens. The white flowers, resembling small single roses, appear in spring and are soon followed by the miniature strawberries: fruits whose fleshy part comes from the flower's basal receptacle that grows around the actual tiny, one-seeded fruits. On this count, botanists refer to strawberries as "accessory fruits."

Horkelia (*Horkelia* spp.). P. 303.

Few nature lovers are acquainted with this shy relative of the cinquefoils or potentillas that was once included with them. This is because these plants are low growing, with an ability to blend into the background of their environment, and because the several white flowers are so small, with only a few open at any one time. The best feature of the plant is the pretty basal clumps of sage scented, ferny, slashed leaves. Details differ as to the toothing or lobing of the leaflets, but all three local species—*H. californica, elata*, and *frondosa* (all three now considered subspecies of *H. californica*)—are closely similar. They are to be sought along the borders of open woods and grasslands.

Sticky Cinquefoil (*Potentilla glandulosa*). P. 306.

The high-elevation species of potentilla are pretty additions to mountain meadows, but the lowland sticky cinquefoil is a widespread but little noticed plant. The basal rosette of leaves is unassuming, and if you look at just the ends of these leaves you get the immediate impression they belong to a strawberry, for the pinnately compound leaves end in three strawberry-shaped leaflets. The slender flowering stalk arises in mid-spring, but it is rather disappointing, as the small cream-color to pale yellow flowers are less showy than those of most rose relatives. Sticky cinquefoil is one of California's great adapters, having different ecological races from coastal bluffs across the foothills, whence they climb into montane meadows, stopping only when they've reached alpine scree above timberline.

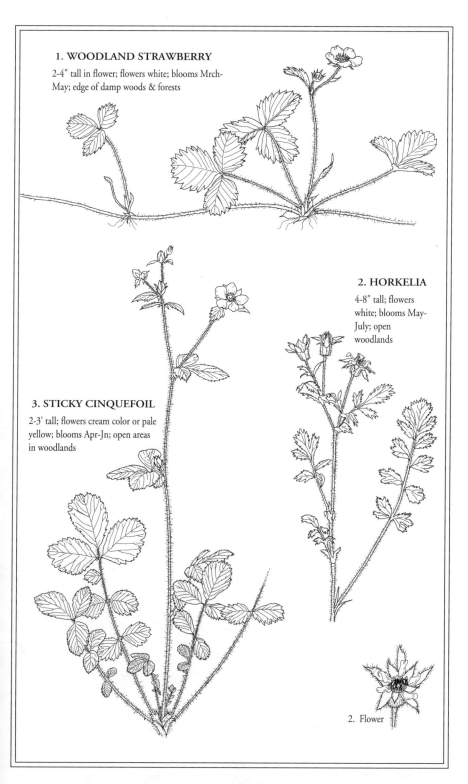

1. WOODLAND STRAWBERRY

2-4" tall in flower; flowers white; blooms Mrch-May; edge of damp woods & forests

2. HORKELIA

4-8" tall; flowers white; blooms May-July; open woodlands

3. STICKY CINQUEFOIL

2-3' tall; flowers cream color or pale yellow; blooms Apr-Jn; open areas in woodlands

2. Flower

SAXIFRAGE FAMILY *(SAXIFRAGACEAE).*

Nonwoody plants with leaves often in basal rosettes and palmately scalloped or lobed. Flowers usually small, often in racemes or panicles; 4 or 5 sepals fused into a cup, 4 or 5 separate petals, 5 or 10 stamens attached to sepal cup rim, usually two nearly separate pistils with ovaries at bottom of sepal cup. Fruit a follicle or capsule.

Alumroot (*Heuchera micrantha*). P. 303.

The heucheras are widely used in horticulture, especially hybrids between the old garden favorite coral bells (*H. sanguinea*) and our own native alumroot. The basal tuft of prettily scalloped leaves often becomes mottled or develops rich purple veining in sunny exposures; this complements the airy panicles of miniature white to pinkish bell-shaped flowers. Alumroot is especially attractive in rock gardens in light shade, where its leaves remain attractive year-round. The name alumroot alludes to an alumlike substance in the root that is said to stop diarrhea.

Woodland Star (*Lithophragma* spp.). P. 304.

The ephemeral blooming period of woodland star is one of the high points of mid-spring, for when those tiny basal rosettes of lobed leaves send up a flowering stalk to a foot or more, the delicate flowers open sequentially from the bottom upwards. Each blossom is a four- or five-pointed white star, the individual petals prettily toothed or fringed in the manner of snow-flakes. After blooming and seeding, woodland stars' tops disappear; food and water are stored in tiny underground tubers through the dry summer. Our two common species are sometimes told apart by leaf color: *L. heterophyllum* (illustrated here) has reddish-purple tinted leaves, while *L. affine* has plain green. In flower, *L. affine* has a vase-shaped sepal cup, while *L. heterophyllum* is characterized by its squared-off, blunt-based sepal cup.

California Saxifrage (*Saxifraga californica*). P. 306.

Unlike our other saxifrage relatives, California saxifrage has entire, unlobed leaves. These are broadly elliptical with short, fuzzy hairs and arranged in tight and tidy rosettes seated against rocks on mossy banks in woodlands. In fact, the name "saxifrage" comes to us from "saxos"—rock—and "fragan"—to break, since so many members of the family grow in crevices and chinks of rocks. California saxifrage sends up a short red-tinted flowering stalk—with a few shorter branches—bearing small, ordinary-looking pretty white flowers that quickly fade and go to seed.

Fringe-cups (*Tellima grandiflora*). P. 307.

When you first see the rosettes of scalloped leaves on fringe-cups, it's hard to believe this is not simply another alumroot, so closely do the leaves match. The proof, of course, is when the flowering stalks appear, for instead of a loose, open panicle with miniature bells, fringe-cups has a dense raceme of cup-shaped flowers held more or less horizontally, with narrow, fringed petals. On some plants petals start out pale green, then deepen to rose-pink, but on others they remain drab. When finally the flowers have finished, the cuplike seed pods open to shed myriad tiny seeds as winds blow the now-brittle flowering stem this way and that. Fringe-cups is easily established in cultivated woodland gardens, but does seed itself enthusiastically.

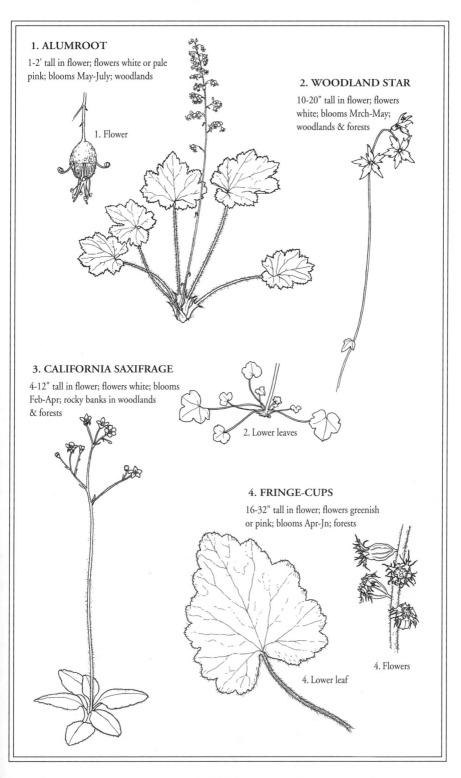

1. ALUMROOT

1-2' tall in flower; flowers white or pale
pink; blooms May-July; woodlands

1. Flower

2. WOODLAND STAR

10-20" tall in flower; flowers
white; blooms Mrch-May;
woodlands & forests

3. CALIFORNIA SAXIFRAGE

4-12" tall in flower; flowers white; blooms
Feb-Apr; rocky banks in woodlands
& forests

2. Lower leaves

4. FRINGE-CUPS

16-32" tall in flower; flowers greenish
or pink; blooms Apr-Jn; forests

4. Flowers

4. Lower leaf

SPURGE FAMILY (EUPHORBIACEAE).

Nonwoody plants or small shrubs, often with copious, caustic milky juice in leaves and stems; leaves simple. Flowers tiny, often greenish, sometimes arranged inside cup-shaped structures, often unisexual. Sepals 5 or lacking, petals missing, 1 to 5 stamens, single pistil with superior, 3-chambered ovary. Fruit a capsule.

Turkey or Dove Mullein (*Eremocarpus setigerus*). P. 302.

Turkey mullein is a true summer annual like the tarweeds, and it is another native that behaves in weedlike fashion. Look for the low, sprawling, grayish mats from mid-summer through early fall along roadsides and paths. The leaves bear harsh hairs but have a pleasant lemony odor when crushed; the flowers—borne in small clusters toward the center of the plant—are greenish and go practically unnoticed. Turkey mullein produces seeds that many wild birds feed on; hence the common name. The common name "mullein" is also applied to a European plant from the figwort family; *Verbascum thapsus* is a striking plant with basal rosettes of wool-covered, pale green leaves and tall spikes of showy, yellow flowers.

Spurges (*Euphorbia* spp.). P. 302.

The genus *Euphorbia* is one of the world's largest, with up to 1,500 species distributed worldwide. Nor do they superficially look alike: South African euphorbias are cactuslike, Mexican species (such as our Christmas flower, the poinsettia) are shrubby, Mediterranean kinds are showy perennials, and California species are prostrate, ground-hugging annuals or taller nonwoody perennials. The genus is characterized by a peculiar floral arrangement: a small cup holds separate male flowers (one stamen each) and a single female flower (consisting of a stalked pistil). The outside of the cup is usually decorated with peculiar, warty nectar-bearing glands that do double duty as petals, often colored white or yellow or spotted with dark purple. In addition, some species surround the floral cups (cyathia) with modified leaves or floral bracts that may be brightly colored (as with poinsettia but not in our native species). A couple of spurges may be familiar garden weeds, both non-native: *E. peplus*, petty spurge, is a pesky upright annual with tiny greenish flower heads scarcely noticed without the aid of a good hand lens; *E.* [*Chamaesyce*] *maculata*, spotted spurge, is a low annual mat with purple spots on the leaves and equally inconspicuous flowers. Occasional in our woodlands is the native spurge *E. crenulata*. This a short-lived perennial with oval leaves and flower cups decorated on the outside with pronged yellow glands (use a hand lens again).

VALERIAN FAMILY (VALERIANACEAE).

Nonwoody plants with simple, opposite leaves. Many small flowers borne in dense spikelike panicles; 5 sepals that enlarge and change form during the fruiting stage, 5 partly fused and slightly irregular petals ending in a spur or sac, 2 stamens, single pistil with inferior ovary. Fruit an achene.

Red Valerian or Jupiter's Beard (*Centranthus ruber*).

Here is a prolific perennial from the Mediterranean that makes itself thoroughly at home on rocky outcroppings around the Bay. Basically a garden escape, red valerian is a pretty wild-flower with striking masses of rose-purple flowers that butterflies find irresistible. A unique feature is revealed by the developing fruits, for at first the sepals on top of the ovary look like a rolled green rim, but gradually they uncurl, finally expanding into intricate white feathers that spirit away the fruits upon winds.

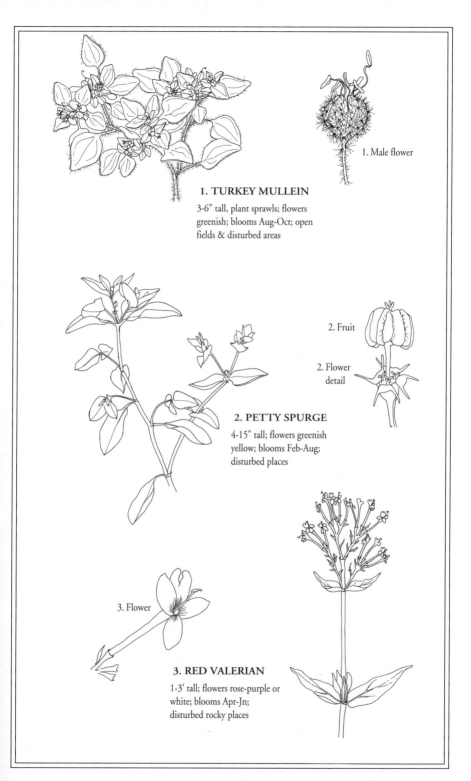

1. Male flower

1. TURKEY MULLEIN

3-6" tall, plant sprawls; flowers
greenish; blooms Aug-Oct; open
fields & disturbed areas

2. Fruit

2. Flower
detail

2. PETTY SPURGE

4-15" tall; flowers greenish
yellow; blooms Feb-Aug;
disturbed places

3. Flower

3. RED VALERIAN

1-3' tall; flowers rose-purple or
white; blooms Apr-Jn;
disturbed rocky places

Native Valerian (*Plectritis* spp.). P. 306.

Our plectritises do not have a well-established common name, yet they're among our most typical woodland annuals in mid-spring. Perhaps it's because of their retiring personalities, for the stems bearing the attractive pairs of pale green leaves fail to deliver showy flowers. Instead, the top is occupied by whorls of tiny whitish to pale pink flowers, each with a minute spur or sac holding nectar. Only when massed together by the dozens do they attract attention, for showier wildflowers such as Chinese houses (*Collinsia heterophylla*), clarkias, and globe tulips overshadow their subdued beauty.

VERBENA FAMILY (VERBENACEAE).

Mostly nonwoody plants with square stems and sometimes scented, opposite leaves. Flowers often in whorls or umbels, slightly irregular; 5 fused sepals, 5 fused petals (sometimes two-lipped), 4 stamens attached at base to petals, single pistil with superior, 4-lobed ovary. Fruit 2 or 4 nutlets.

Common Vervain (*Verbena lasiostachys*). P. 308.

Common vervain is a striking yet weedy-looking perennial, thriving on open rocky slopes and sides of dry arroyos. On first acquaintance you're liable to think it a mint relative, for it has two key mint family features—square stems and opposite leaves. But there is no minty odor. Actually, the verbenas and mints are close relatives, and even though the flowers are seldom conspicuously two-lipped (as are most mints) they show a close relationship in their disposition and number of parts. Common vervain bears a profusion of pretty but small pale purple flowers, best appreciated at close range. The many-branched stems create an open architecture for the whole plant.

VIOLET FAMILY (VIOLACEAE).

Nonwoody plants with simple to deeply divided leaves that have stipules. Flowers usually showy, irregular, and spurred: 5 separate sepals; 5 petals, 2 above and 3 below, with the middle lower petal often enlarged or marked with lines; 5 partly fused stamens at entrance to spur; single pistil with superior, 3-sided ovary. Fruit an explosive capsule.

Wild Violets (*Viola* spp.). P. 308.

The violets include some of our prettiest spring wildflowers. We have a wide range of different kinds, each with its own special charm. Dog violet (*V. adunca*) slowly creeps over the ground among short grasses on bluffs facing the Bay and has blue to violet flowers. Two pansylike violets—*V. pedunculata* (pictured) and *V. douglasii*—frequent the grasses of rolling hills and bear unusually large, butterscotch-yellow blossoms with brown markings. The first has the usual heart-shaped leaves of so many violets, while the second has leaves deeply slashed into many narrow segments. Both act like our native bulbs; they dry up and go dormant to deeply seated roots in summer and fall. These charmers are fast disappearing to urbanization, but fortunately the more remote reaches of our parks and uplands along the Mines Road still are covered with hundreds of yellow pansies in spring. Finally, we have a trio of yellow-flowered woodland violets: *V. sempervirens* (redwood violet) grows as low, sprawling mats with nearly round leaves; *V. purpurea* (pine or oak violet) has narrow leaves and grows as clumps in rocky places; and *V. sheltonii* (Shelton's violet) has distinctive, deeply slashed leaves.

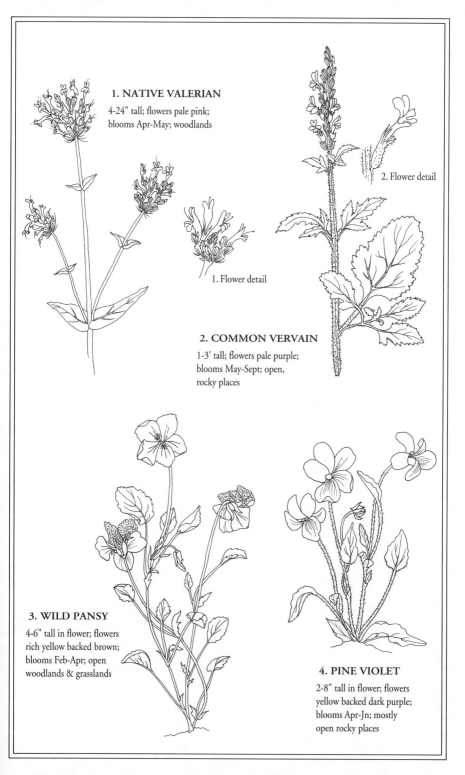

1. NATIVE VALERIAN
4-24" tall; flowers pale pink;
blooms Apr-May; woodlands

2. Flower detail

1. Flower detail

2. COMMON VERVAIN

1-3' tall; flowers pale purple;
blooms May-Sept; open,
rocky places

3. WILD PANSY

4-6" tall in flower; flowers
rich yellow backed brown;
blooms Feb-Apr; open
woodlands & grasslands

4. PINE VIOLET

2-8" tall in flower; flowers
yellow backed dark purple;
blooms Apr-Jn; mostly
open rocky places

WATERLEAF FAMILY (HYDROPHYLLACEAE).

Small shrubs or nonwoody plants with variable leaves. Flowers borne in coiled clusters that unroll as buds open; 5 partly fused sepals, 5 partly fused petals, 5 stamens fused at bases to petals, single pistil with 2-chambered superior ovary and 2-branched style. Fruit a capsule.

Whispering Bells (*Emmenanthe penduliflora*). P. 302.

Whispering bells is a special annual that is seldom seen unless there has been a chaparral burn. Afterward, they come up by the hundreds. These are special flowers in every respect. The prettily pinnately lobed leaves are scented when dry; fresh, they complement the pale yellow bell-like flowers that later nod as they fade. When flowers dry they rattle about in breezes, giving the common name "whispering bells." Were it not for the flower color and nodding disposition of flowers, it would be easy to mistake whispering bells for one of the phacelias.

Eucrypta (*Eucrypta chrysanthemifolia*).

"Eucrypta"—the name—means truly hidden in Greek, but whether for the inconspicuous white flowers or the rarity of the plants might be disputed. Actually, the plants are not rare when you know that eucrypta's special niche is that of a fire follower; thereafter the seeds germinate in great profusion, the plants smothering whole hillsides where chaparral shrubs once stood. Though a close relative to the nemophilas, eucrypta's chrysanthemumlike foliage is pleasantly scented, and the sepals have no down-turned lobes between them.

Baby-blue-eyes (*Nemophila menziesii*).

Well-known as a garden annual, baby-blue-eyes was introduced to its original homeland by English horticulturists who early recognized the charm of this favorite annual wildflower. Look for fields of sky blue saucers in mid-spring, often in company with California poppies and popcorn flower. Most have blue petals with a small white center, but the lovely subspecies *atomaria* has palest lavender-blue flowers lined with purple stripes and sprinkled with purple dots. The genus *Nemophila* also contains a few smaller-flowered species with pure white to palest purple petals. These seek shade, as the genus name suggests: "nemos" is Greek for woodland and "philos" means lover. Nemophilas all bear the mark of their clan: tiny, down-pointing earlobelike projections between the sepals (a hand lens is recommended).

California Phacelia (*Phacelia californica*). P. 306.

California phacelia is one of our most easily recognized phacelias and is habitat specific, preferring rocky outcrops and cliffs in full sun. There, the low rosettes of unusual leaves distinguish themselves by their pinnately engraved veins, silvery to whitish adpressed hairs, and coarsely pinnately divided leaves. The terminal segment is longest and most obvious. In bloom California phacelia may be disappointing, for the fuzzy caterpillarlike, coiled flower clusters open to dull, off-white blossoms, distinguished at close range by long, protruding stamens. The same species from coastal promontories differs by having tidier-looking flowers in various shades of purple, and the plants are more compact and altogether more attractive. The coastal form does well in rock gardens with excellent drainage. A related species with similar but greener leaves and taller flowering stalks is bristly phacelia, *P. nemoralis*, another distinctive scree-loving plant found near the summit of Mt. Diablo.

3. BABY-BLUE-EYES

2-4" tall; flowers usually blue w/ white center;
blooms Mrch-May; grasslands, open woodlands

2. EUCRYPTA

8-20" tall; flowers white; blooms
Apr-May; rocky or grassy places,
especially after burns

4. Lower leaf

1. WHISPERING BELLS

4-20" tall; flowers pale yellow;
blooms Apr-July; rocky scree or
burned areas

4. CALIFORNIA PHACELIA

4-32" tall; flowers Apr-Jn; flowers dirty
white; rocky places

Caterpillar Flowers (*Phacelia* spp.). P. 305, 306.

Most phacelias, and they number over eighty species for the state, resemble wooly to fuzzy-haired caterpillars when their flowers are coiled in bud. Of the many types, California phacelia is perennial, but many others are spring-flowering annuals. Of these, several closely resemble one another, with delicate, almost lacy, fernlike leaves and modest clusters of white to pale purple flowers. Perhaps the most unusual is the nettle phacelia (*P. malvifolia*), which is occasional in our area. It has palmately lobed, mallowlike leaves covered with coarse, stinging hairs that like many other species cause contact dermatitis for some people (though most are immune). Other typical local phacelias include two with fernlike, dissected leaves: *P. distans* (illustrated) has whitish flowers while *P. tanacetifolia* (tansy phacelia) has purplish flowers. Both are further distinguished by long, protruding stamens. Two small annuals with more ordinary leaves are *P. divaricata*, with showy blue-purple flowers and undivided leaves, and *P. breweri*, with tiny blue flowers and lobed leaves. Finally, there's a close phacelia relative: western waterleaf (*Hydrophyllum occidentale*). Scattered under dense shrubberies or low trees near the top of Mt. Diablo, waterleaf has distinctive yellow-green splotches on its leaves and pale purple flowers in tight, headlike clusters.

The spotted leaves are said to look like water droplets, hence the common name "waterleaf" and the scientific name "Hydrophyllum."

Fiesta Flower (*Pholistoma auritum*). P. 306.

Take baby-blue-eyes and add the ability to climb by clinging stems and you've got a fiesta flower or pholistoma. Like most nemophilas, fiesta flower is a lover of lightly shaded banks and edges of woodlands, where its weak-stemmed, sprawling branches find support by grasping as they climb. Even when picked you can easily demonstrate the tenacity of the stems by placing one on your clothing, where the plant will readily cling. The hand lens reveals the mechanism for this ability; the stems bear vertical rows of backwardly pointing pricklets with grappling-hook tips. In mid-April, fiesta flower is covered with inch-broad deep blue, shallowly bowl-shaped blossoms centered white. If it weren't for the awkwardness of the growth, fiesta flower would be as charming as baby-blue-eyes in a woodland garden. A second less-common species is *P. membranaceum*, an annual with shorter branches, smaller leaves, and considerably smaller, modest white flowers.

1. CATERPILLAR FLOWER

4-32" tall; flowers pale bluish; blooms
Mrch-Jn; open woodlands & rocky slopes

3. FIESTA FLOWER

1-3½' tall; flowers purple;
blooms Mrch-May;
woodlands

1. Leaf

3. Stem

2. WESTERN WATERLEAF

8-24" tall; flowers pale purple; blooms
May-Jn; open woodlands

Section III
References and Resources

Appendix A: Places to Go See the Plants
The Regional Parks

Below you'll find a listing of the major regional parks in Alameda and Contra Costa counties, the geographical region covered by this book. For each park there is a short description of park highlights, the vegetation, and unusual or special plants. Keep in mind that most shoreline or bayside parks are not included.

Park descriptions are followed by a map that keys each park and its major trails to a master plant list. Symbols on the map are as intuitive as possible, for easy reference to the plant list.

For lists of the parks' species of trees, shrubs, vines, and wildflowers in alphabetical order, see the Master Plant Location List, Appendix B.

The Parks

BLACK DIAMOND MINES REGIONAL PARK

To get there: Take Hwy. 4 to the Somersville Road exit, and drive south on Somersville Road to the entrance.

Highlights: The park features historical coal mines as well as wide variety of plant life. There are 34 miles of excellent trails.

Plants: The trails go through grasslands, foothill woodland, mixed evergreen forest, chaparral, riparian vegetation, and exotic plantings. The last include black locust, pepper tree, almonds, and Chinese tree of heaven. Look for excellent display of spring wildflowers in years with ample rainfall.

Notable: This is the northernmost area for Coulter pine, black sage, and desert olive. Such Mt. Diablo endemics as Mt. Diablo globe tulip, Mt. Diablo manzanita, and Mt. Diablo sunflower also occur here. Both bush poppy and prickly poppy are also notable, since they're uncommon in our area. Other special shrubs include flowering ash, golden fleece, matchstick, pitcher sage, Fremont bush mallow, and hopbush. Highlights among uncommon wildflowers are scarlet bugler, rush-rose, wild tobacco, golden eardrops, and padre shooting star.

BRIONES REGIONAL PARK

To get there: The park is between the towns of Orinda, Lafayette, Martinez, and Pleasant Hill. The several entrance points include Bear Creek Road, Alhambra Valley Road, Reliez Valley Road, and Pleasant Hill Road.

Highlights: Briones was formerly part of a couple of large ranches, with cattle ranching historically important. There are many miles of excellent and easy walking trails.

Plants: Vegetation includes rolling grasslands, oak woodland, and mixed evergreen forest. Much of the grassland has been disturbed by grazing, with many weedy species.

Notable: The several species of oaks—including valley, Garry, blue, and coast live—are well represented. This is also a good place to see such shade lovers as pipevine, giant trillium, and giant chain fern (*Woodwardia fimbriata*), and spotted coral root. Interesting shrubs include service berry and bitter cherry. Seasonal wildflowers are represented by wild snapdragon, yellow mariposa, ookow, canchalagua, golden aster, wind poppy, red ribbons clarkia, and elegant clarkia.

CARQUINEZ STRAITS REGIONAL PARK

To get there: Along the delta between Crockett and Martinez. Entry is from the Crockett side along Carquinez Scenic Drive at the Bull Valley Staging Area, or from Martinez along Carquinez Scenic Drive at the Carquinez Strait East Staging Area.

Highlights: The parkland provides a gateway to the delta area, with fine views from the 750-foot-high ridge, and looks over marshes and distant mountains. The park is still being developed.

Plants: The vegetation consists of rolling grasslands, wooded canyons, eucalyptus groves, and river shoreline.

Notable: Several coastal plants are to be found here, such as horkelia, blueblossom ceanothus, and toadflax. Other interesting plants include the pipevine, hopbush, and California tea.

ANTHONY CHABOT REGIONAL PARK

To get there: In the hills east of Oakland and San Leandro. There are multiple entry points for the park, including several along Redwood Road (35th Avenue in Oakland becomes Redwood Road past Skyline Boulevard). Other entry points include the Grass Valley staging area from Skyline Boulevard and the Lake Chabot Marina from Chabot Road.

Highlights: The most visited part of the park centers around the artificial lake named for Anthony Chabot, but there are many miles of windy and convoluted trails, and the new Skyline National Ridge Trail passes through. The park was formerly part of large ranches.

Plants: Vegetation includes old eucalyptus groves—some cut down—as well as mixed-evergreen forest, riparian woodland, and coastal scrub, with abundant coyote bush. The latter has invaded several areas that were formerly grasslands for grazing. Even though the vegetation is disturbed in many places, the park is home to many interesting plants that are uncommon to rare for our area.

Notable: Several interesting shrubs, including service berry, ninebark, twinberry honey-suckle, bitter cherry, brittle-leaf manzanita, huckleberry, and the rare western leatherwood. Among the unusual trees are California bayberry and interior live oak, the latter normally farther inland. Several unusual water or marsh plants include arrowhead, *Echinodorus*, rosilla, St. Johnswort, leopard lily, and water knotweed. Uncommon forest plants are baneberry, angelica, red columbine, giant trillium, and western bleeding heart.

CONTRA LOMA REGIONAL PARK

To get there: From Hwy. 4 in Antioch turn onto Lone Tree Way, then onto James Donlon Boulevard. The main entrance is left from this boulevard and takes you across the park into Black Diamond Mines Regional Park.

Highlights: Contra Loma shares its southern border with Black Diamond Mines Regional Park; several trails continue through both. Much of the park is occupied by a central reservoir, with beaches and swimming.

Plants: Vegetation is grassland on rolling hills, with some scattered oaks. In years of ample rain, good wildflower shows occur in these grasslands.

Notable: There are few unusual or rare plants, but some of the less common wildflowers to be found here include balsamroot, blazing star, monolopia, and tansy-leafed phacelia.

DEL VALLE REGIONAL RECREATION AREA

To get there: From Livermore, take Livermore Avenue south out of town. At a bend Livermore Avenue becomes Tesla Road. Continue on Tesla Road to Mines Road, turn right, and follow signs to Del Valle Road.

Highlights: Del Valle, Sunol, Ohlone Camp, and Mission Peak all abutt one another. Del Valle is centered entirely around a reservoir created by damming Arroyo del Valle, one of the largest creeks in the area. There are few trails; most follow the reservoir.

Plants: The vegetation consists of grasslands dotted with oaks; oak and foothill woodland on steeper slopes, with digger pines; and chaparral.

Notable: California juniper is seen in some places with digger pines; the stream bottom is home to some fine sycamores. California sunflower is a giant stream follower. Unusual wildflowers include the spectacular blazing star *Mentzelia laevicaulis,* the white fiesta flower *Pholistoma membranaceum,* and wind poppy.

DIABLO FOOTHILLS (AND CASTLE ROCK) REGIONAL PARKS

To get there: Take Ignacio Boulevard from Hwy. 680; turn right onto Walnut Boulevard, then jog right at the intersection with Oak Grove Road. At the first fork, North Gate Road is to the left; stay right on Castle Rock Road and follow it to its end.

Highlights: Once part of a large Mexican land grant for a ranch, the park includes the adjacent Castle Rock Recreation Area. The several trails wind through rugged hills and the canyon of Pine Creek. The east side is bounded by the rugged Castle Rocks—picturesque rocks with good bird eyries—and Mt. Diablo State Park; the west side is bordered by Shell Ridge Open Space, with a series of razor-backed hills. From Pine Canyon you can continue on to Pine Pond in the Mt. Diablo State Park itself.

Plants: Oak savannah and woodland border rocky slopes with chaparral. The canyon bottom along Pine Creek has excellent riparian woodland. Spring flowers are quite good here, but there is also the element of weedy species immediately along Pine Canyon trail.

Notable: Several important riparian trees occur along Pine Creek, including big leaf maple, box elder, walnut, and Fremont cottonwood. Other interesting trees include occasional madrones and also California black and valley oaks. Interesting shrubs are oso berry, bitter cherry, hopbush, and squawbush. Among the several bulbs are harvest brodiaea, yellow mariposa, Mt. Diablo globe tulip, and ookow. In addition you may find such riparian wildflowers as wild licorice, rosilla, and an isolated stand of hummingbird sage. The latter is rare in our area.

GARIN REGIONAL PARK

To get there: East of Hayward and Union City. From Hwy. 580, take Foothill Boulevard east, and turn left onto Mission Boulevard. The main entrance is left on Garin Avenue to its end. Further south on Mission Boulevard, Tamarack Drive takes you into adjacent Dry Creek Pioneer Regional Park.

Highlights: Park headquarters houses displays of old ranch equipment and an old barn. There are twenty miles of winding trails, some around hills to nearly 1,000 feet elevation.

There's also a pond for fishing.

Plants: Vegetation includes oak woodland, grass-covered hills, and some good stands of riparian vegetation, including sycamores.

Notable: Among the riparian and wet-growing plants are creek dogwood, ninebark, water parsnip, elk clover, rosilla, and water knotweed. An unusual woodland shrub for our area is choke-cherry. Wildflowers of interest include white fairy lantern, California lomatium, balsamroot, and fiesta flower.

HUCKLEBERRY REGIONAL PRESERVE

To get there: In the Berkeley-Oakland hills, turn left onto Skyline Boulevard from Grizzly Peak Boulevard, go about one-half mile, and park in the pull-out on the east side of the road.

Highlights: This preserve is just what it says: a place to preserve unusual plant life in its natural condition. A system of three trails go to different levels into the canyon or along steep hillsides. All three are well worth the time for the many unusual and beautiful plants in a mixed-evergreen forest setting.

Plants: Besides mixed-evergreen forest dominated by coast live oak, madrone, and extensive bays, there's riparian vegetation along the canyon bottom, localized springs, and north coastal scrub on exposed slopes.

Notable: This area is one of the best to explore for plants that normally grow in more coastal or wet situations, and features such rare shrubs as western leatherwood and pallid or Alameda manzanita. Actually, the manzanitas here show hybridization between the burl-forming brittle-leaf manzanita and the treelike pallid manzanita. Other notable shrubs include shrub chinquapin, evergreen huckleberry, silk tassel bush, oso berry, creek dogwood, and the small trees as hazelnut and California bayberry. Woodland wildflowers—many typical of the redwood belt—include fringe-cups, alumroot, false Solomon's seal, fairy bells, star flower, Douglas iris, red columbine, coast trillium, baneberry, and the orchid known as spotted coral root.

LAS TRAMPAS REGIONAL PARK

To get there: West of Danville and San Ramon; east of Castro Valley. Take Crow Canyon Road west from Hwy. 680, turn north onto Bollinger Canyon Road, and follow it to its end.

Highlights: A rugged park with ridges to 2,000 feet elevation, and deep valleys. Heavily faulted terrain with excellent wildlife and several trails.

Plants: Some grassland in the valleys with riparian woodland, but mostly chaparral on hot slopes and oak woodlands and mixed-evergreen forest on cool slopes.

Notable: In oak woodlands, look for Garry oak, which is rare in our area. Notable shrubs include service berry, silk tassel bush, and bitter cherry. The curious California pipevine is here. Beautiful spring wildflowers include white fairy lantern, red ribbons clarkia, tuberous skullcap, and *Collinsia bartsiaefolia*, an unusual kind of Chinese houses. Flame and wind poppies are fire followers that occur here.

MISSION PEAK REGIONAL PARK

To get there: East of Fremont. Take the northern Mission Boulevard exit from Hwy. 680, go past the entrance to Ohlone College, and take the next left to the trailheads, or continue down Mission Boulevard to Stanford Avenue, and go left and follow it to its end.

Highlights: Besides being contiguous with Sunol Regional Wilderness, the park is a spectacular place to hike the several trails, especially to the top of 2,500-foot Mission Peak with its sweeping vistas. Hang gliders also are common sights here.

Plants: The park is mostly home to oak woodlands and disturbed grasslands, with some local streamside vegetation in places.

MORGAN TERRITORY

To get there: East of Mt. Diablo, southeast of Concord and Clayton, northeast of Danville, and north of Livermore. Take Morgan Territory Road north from Hwy. 580 by Livermore or, from Clayton, go south on Marsh Creek Road to its split with Morgan Territory Road, and turn right onto the latter.

Highlights: Morgan Territory was previously the homeland to the Volvon peoples, and later was developed into a ranch by Jeremiah Morgan. Morgan Territory is a complex of steeply folded sandstone hills and ridges east of Mt. Diablo, with spectacular views on clear days of the Sierra, Mt. Diablo, and Mt. St. Helena. There is an ample trail system.

Plants: Riparian woodland in deep canyons, but the best features are the dense oak or oak and bay woodlands, with chaparral on the steep, south-facing slopes. Also many areas of open grasslands, with considerable numbers of weedy species in places.

Notable: Riparian plants include sycamores, Fremont cottonwood, and wild grapes. Oso berry, chaparral pea, choke-cherry, hopbush, and squawbush are among the more unusual shrubs. Most of the area's oaks are also here, including canyon live, coast live, interior live, blue, valley, and California black oaks. Several spring bulbs are of note: harvest and white brodiaeas, white lava and single-leaf onions, checker lily, and Mt. Diablo globe tulip. The many beautiful wildflowers include red ribbons clarkia, wild pansy, Brewer's flax, Indian pink, foothill meadow rue, and such rock dwellers as bitterroot and annual stonecrop.

MT. DIABLO STATE PARK

To get there: The two major roads into the park are North Gate Road—accessed from Walnut Boulevard off Ignacio Valley Boulevard; and South Gate Road—accessed from Hwy. 680 by taking the Diablo Boulevard exit. Other access is through various trailheads (see map #2), such as from Regency Drive, Clayton for Back and Donner canyons; or from the end of Mitchell Canyon Road in Clayton.

Highlights: This is the preeminent park of our area, with the top over 3,800 feet in elevation. You'll see spectacular views of northern California as far as the Sierra on clear days. An excellent new museum is located at the summit. There are several camping areas, and many miles of fine trails with a broad range of habitat and terrain. Several major canyons have intermittent to perennial streams.

Plants: The plant communities here range the gamut from riparian woodlands to oak savannah and woodlands; from hard chaparral to soft chaparral; from mixed woodlands to open grasslands; and they include some excellent rock-adapted vegetation. Also a few areas of serpentinite rocks, with specialized endemic plants.

Notable: Besides the wide array of oaks, riparian trees, bays, madrones, and digger pines, Coulter pine is found in several good stands, and California juniper is found in several sites. Endemic to the area around the mountain is Mt. Diablo sunflower, Mt. Diablo globe tulip, Mt. Diablo manzanita, and Mt. Diablo jewel flower. Burns provide special fire followers, such as Fremont bush mallow, flame and wind poppies, whispering bells, and golden eardrops.

Spring wildflowers and bulbs are too numerous to list, but two unusual bulbs beyond their normal range are lilac mariposa and Hooker's onion. Special rock-dwelling plants include bitterroot, red rock penstemon, sulfur buckwheat, and sickle-leaf onion.

OHLONE REGIONAL WILDERNESS

To get there: This is a remote park, placed between Sunol Regional Wilderness and Del Valle Recreational Area. You have to hike seven miles up Alameda Creek from Sunol to get there; there is no road access.

Highlights: This area offers long day hikes and group overnight camping by advance reservations only.

Plants: Plant life is similar to the descriptions for Del Valle and Sunol.

Notable: The area is well worth exploring for several unusual or uncommon plants. Special shrubs include service berry, rabbitbrush (rare in our area), matchstick, and choke-cherry. Scree plants include annual harebell, spine flowers, *Collinsia bartsiaefolia*, sulfur buckwheat, bitterroot, Lindley's blazing star, tuberous skullcap, common jewel flower, and pine and Shelton's violets.

REDWOOD REGIONAL PARK (INCLUDING ROBERTS)

To get there: In the hills east of Oakland. Several entry points along Skyline Boulevard; also from Redwood Road (the continuation of 35th Avenue east of Skyline) and Pinehurst Road, which splits from Redwood Road and continues on to Moraga.

Highlights: Highly accessible area, with many picnic facilities, including Roberts Recreation Area. Just below is Joaquin Miller Park and Woodminster. Several miles of trails, including part of the new Skyline Ridge Trail system. Some parts are highly disturbed, but there is also much to see here.

Plants: This is one of our few areas of well-developed redwood forest, with many of the associated understory plants, such as huckleberry, thimbleberry, and hazelnut. Also good areas of riparian woodland, and coastal scrub, plus naturalized groves of Monterey pine and some fine serpentinite slopes along the southwest side, off Skyline Boulevard.

Notable: Besides such riparian trees as bigleaf maple, box elder, white alder, and California bayberry, there are stream-following shrubs, such as ninebark and oso berry. Western leather-wood, one of our rarest shrubs, is also here. Notable redwood forest wildflowers include baneberry, elk clover, wild ginger, bead lily, trilliums, Douglas iris, western bleeding heart, violets, alumroot, and fringe-cups. Mostly coastal species include golden eggs, footsteps-to-spring, dog violet, and pearly everlasting.

SIBLEY VOLCANIC REGIONAL PARK

To get there: In the hills behind Oakland. Turn left onto Skyline Boulevard from Grizzly Peak Road. Within less than one-quarter mile, turn left into the park. The road takes you to a visitor's center, or you may continue to the summit of Round Top at 1,760 feet.

Highlights: This small park highlights an eroded-away view into an ancient volcanic plug. Geologic formations include basaltic dikes, tuff-breccias, lava flows, red-baked cinder piles, and a major volcanic vent.

Plants: Most of the area is disturbed, and it once was covered with blue gum eucalyptus forest. Many of the gums died in the 1972 freeze. There is a grove of Monterey pines on top of Round Top.

Notable: Few unique or unusual plants occur here, but if you're hiking in adjacent Huckleberry Preserve you might want to check out this area for its geology. One rare plant—the coneflower—has been spotted in this park.

SUNOL REGIONAL WILDERNESS

To get there: East of Fremont and south of Pleasanton. From Hwy. 680 south of Pleasanton, take the Calaveras Road exit, turn left, then left again on Geary Road and into the park.

Highlights: This urban "wilderness" consists of rugged country around the headwaters of Alameda Creek: the largest stream in our area. Rugged cliffs and steep hills to over 2,000 feet add to the scene. Several trails, backpacking campsites, and a nature area are included here. Sunol adjoins Ohlone Wilderness and Del Valle Recreational Area.

Plants: Plant communities include fine riparian corridors, oak woodlands, grasslands, and chaparral. This is a good place to see a variety of large-specimen trees.

Notable: Notable riparian plants include large sycamores, durango root, Indian hemp, wild licorice, and California sunflower. Uncommon shrubs are choke-cherry, service berry, and Douglas's bush senecio. The last reaches its northern limits around here. Colorful wildflowers include Brewer's butterwort, Indian pink, pine violet, common stonecrop, tuberous skullcap, fiesta flower, wild snapdragon, and Lindley's blazing star.

TILDEN REGIONAL PARK

To get there: Encompassing part of Wildcat Canyon in the Berkeley hills, with multiple entry points. Grizzly Peak Boulevard and Wildcat Canyon Road in Berkeley offer several points of entry. Also, Shasta Road and Golf Course Drive (a continuation of Centennial) enter the park from Grizzly Peak Boulevard. (Or enter the park from the east side via Wildcat Canyon Road from Orinda.)

Highlights: Tilden offers a diversity of activities and trails, including the nature area near Jewel Lake, sweeping vistas of the Bay Area from Inspiration Point, the steam trains, a golf course, Lake Anza, and a fine Botanical Garden devoted exclusively to California native plants. The latter is located by the junction of South Park Drive and Wildcat Canyon Road.

Plants: Much of the park is highly disturbed, with many man-planted blue gum eucalyptus and Monterey pines. Despite this, there are fine natural areas, including rolling grasslands (though many of these are being taken over with coyote bush), coastal scrub, riparian vegetation, localized redwoods, and mixed-evergreen forest.

Notable: Riparian or wet-growing plants include bigleaf maple, white alder, California bayberry, red elderberry, blueblossom, creek dogwood, western burning bush and twinberry honeysuckle. Douglas's baccharis, yellow-eyed grass, leopard lily, and brook orchid grow on adjacent EBMUD land. Associates of redwood and other damp forests are baneberry, wild ginger, spotted coral root, Douglas iris, checker lily, trillium, meadow rue, star flower, and false Solomon's seal. Other interesting flowers include white mariposa, ookow, golden aster, cow parsnip, California skullcap, and angelica. Also there are at least two stands of the rare western leatherwood, with some quite sizeable specimens.

WILDCAT CANYON REGIONAL PARK

To get there: East of Richmond and west of El Sobrante. Take the McBryde exit from I-80 to its end to reach the main entrance. Access is also possible from Clark Road, which branches off San Pablo Dam Road, or by trail from the Jewel Lake area in contiguous Tilden Regional Park.

Highlights: Wildcat was once home to groups of Native Americans. Later, squatters' struggles and water wars scarred the landscape. Included within the boundaries is Alvarado Park, with several recreational facilities. Notable there are old stonework buildings with rustic architecture.

Plants: The dominant communities are riparian woodland; mixed-evergreen forests of bay, madrone, and coast live oak; and coastal scrub. There are also grassy areas, some even with the original native bunchgrasses.

Notable: There are only a few unusual or rare plants here. Special riparian plants include the red elderberry and scarlet monkeyflower, while blueblossom ceanothus and California pipevine are other uncommon plants. The most noteworthy plant here is coast iris—*Iris longipetala*—which is very rare in our area.

Other Notable Locales to Visit

While this book is designed to cover much of the East Bay's parklands, other areas are noteworthy, too. I'll detail four of these below, although other areas are cited in the master plant list that follows.

Joaquin Miller Park.

This popular recreational area is a city park just below Redwood Regional Park in the Oakland hills. There are serpentine outcrops here, as well as deep shaded gullies. Notable wildflowers to look for include Indian pink, Henderson's shooting stars, Oakland star tulip, trillium, checker lily, and false Solomon's seal.

Antioch Dunes Preserve

This preserve requires special permission to enter and is the last remnant of a highly unique plant community, featuring sand dunes along the edge of the delta. These dunes are home to several unusual plants, including the beautiful rare and endangered Antioch dunes evening primrose—*Oenothera deltoides howellii*—and dunes wallflower—*Erysimum capitatum angustatum*.

Corral Hollow

This area and adjacent Livermore—in places where development hasn't overwhelmed the natural environment—are excellent repositories of unusual habitats and plants. The rare wetlands known as vernal pools still occur adjacent to Livermore.

The Corral Hollow area itself represents an especially dry, hot habitat comparable to deserts to our south, and with many desert species extending their ranges this far north. Expect to see such typical desert species as thistle sage, wavy onion, *Calyptridium monandrum*, Wright's buckwheat, Coulter's dandelion, and a form of birdcage evening primrose.

Inland foothill wildflowers unusual to our area include stink bells, white fiesta flower, lilac jewel flower, monolopia, evening snow, and California wild flax.

To get there, go out Hwy. 580 to Livermore; turn south on Livermore Avenue. At a bend, the road becomes Tesla Road. Continue out Tesla beyond the turn-off to Mines Road for a few miles. This road now becomes Corral Hollow Road and extends across the Alameda County line before finally rejoining Hwy. 580 to the southeast.

Mines Road

The Mines Road is one of the very best roads for seeing a wide range of colorful spring wildflowers. It stretches for nearly fifty miles from its beginning just off Tesla Road (see above) to the summit of Mt. Hamilton. Plant communities en route include rolling grasslands, oak and foothill woodland with digger pines and California junipers, riparian woodland with sycamores, chaparral, and—near the top of Mt. Hamilton—ponderosa-Coulter pine forest. Be warned that access into the surrounding countryside is limited in most areas, because of sturdy barbed wire fences and "no trespassing" signs.

Cedar Mountain is a serpentine area of particular importance just off Mines Road, where the "cedars" are actually Sargent's cypresses. Unfortunately, Cedar Mountain is off limits without special permission. You'll find many rare and unusual plants from here on the master list.

Among the special shrubs along this road are big-berry manzanita, choke-cherry, service berry, mulefat, desert olive, gaping penstemon, broom scale, bush mallow, golden currant, oak-leaf gooseberry, and white-thorn. Beautiful bulbs include white fairy lantern, yellow mariposa, and white mariposa. Showy annual wildflowers include mickey mouse monkeyflower, Kellogg's monkeyflower, wild tobacco, Brewer's phacelia, elegant clarkia, spine flower, and elegant clarkia. Perennial wildflowers to look for are Jimson weed, California and royal larkspurs, padre shooting stars, Wright's buckwheat, isopyrum, thick-leaf lotus, Brewer's butterwort, Douglas's and pine violets, and wild pansy.

Symbols for Map Locations

Albany Hill	**AlbHll**	Morgan Territory	**MrgTrt**
Antioch Dunes	**AntDns**	Mount Diablo State Park	**MtDbl**
Arroyo Mocho (Mines Road)	**ArryMch**	Arroyo del Cerro	**ArryCrr**
Black Diamond Mines	**BlkDmnd**	Back Canyon	**BckCyn**
Briones Regional	**Park Brns**	Black Hills	**BlkHll**
Browns Island	**BrwnIsl**	Blackberry Spring	**BlkbrSpr**
Byron area (including		Cave Point	**CvPt**
hot springs)	**Byrn**	Curry Canyon	**CurrCyn**
Canyon (town of)	**Cyn**	Dan Cook Canyon	**DnCkCyn**
Carquinez Straits	**CrqzStrt**	Deer Flat	**DrFlt**
Cedar Mountain (Mines Road)	**CdrMt**	Deer Ridge	**DrRdg**
Chabot Regional Park	**Chbt**	Devil's Elbow	**DvlsElbw**
Contra Loma Regional Park	**CntrLm**	Devil's Pulpit	**DvlsPlpt**
Corral Hollow	**CrrlHw**	Donner Canyon	**DnnrCyn**
Del Valle Regional Park	**DlVll**	Eagle Peak	**EglPk**
Diablo Foothills	**DblFthl**	Eagle Ridge	**EglRdg**
Dimond Park, Oakland	**DmndPrk**	Emmons Canyon	**EmnsCyn**
Dry Creek-Pioneer	**DrCrk**	Fossil Ridge	**FsslRdg**
EBMUD land	**Ebmd**	Hetherington Loop	**HthrngtnLp**
El Sobrante Ridge	**ElSbrnt**	Hidden Pond	**HddnPnd**
Fairmont Ridge	**FrmntRdg**	Juniper Camp	**JnpCmp**
Flicker Ridge	**FlkrRdg**	Knobcone Ridge	**KnbcnRdg**
Franklin Canyon	**FrklnCyn**	Maple Spring	**MplSpr**
Garin Regional Park	**Grn**	Mitchell Canyon	**MtchCyn**
Gateway	**Gtwy**	Murchio Gap	**MrchGp**
Grizzly Peak Boulevard	**GrzPk**	Murchio Gate	**MrchGt**
Huckleberry Trail	**Hklbr**	North Gate Road	**NrthGtRd**
Joaquin Miller	**JqnMllr**	North Peak	**NPk**
Lake Temescal	**LkTmscl**	Oyster Ridge	**OystrRdg**
Las Trampas Regional Park	**Trmps**	Pine Canyon	**PnCyn**
Livermore area	**Lvrm**	Pine Pond	**PnPnd**
Los Vaqueros	**Vqrs**	Prospector's Gap	**PrsptrGp**
Marsh Creek area	**MrshCrk**	Ransom Point	**RnsmPt**
Mines Road	**MnsRd**	Rhine Canyon	**RhnCyn**
Mission Peak	**MssnPk**	Rock City	**RckCty**
		Roundtop	**Rndtp**
		Russelman Park	**RslmnPrk**

South Gate Road	SthGtRd
Sulphur Springs	SlphrSpr
summit of mountain (includes Fire Interpretive Trail)	smt
Sycamore Canyon	SycCyn
Turtle Rock	TrtlRck
Walker Ridge	WlkRdg
Waterfall Trail	WtrflTrl
White Canyon	WhtCyn
Willow Springs	WllwSpr
Windy Point	WndyPt
Oakland hills	OkHlls
Ohlone Regional Wilderness	Ohln
Pleasanton Ridge	PlsntRdg
Point Molate	PtMlt
Point Pinole	PtPnl
Redwood Regional Park	Rdwd
San Leandro Reservoir	SLRes
San Pablo Reservoir	SPRes
Shell Ridge	ShlRdg
Sibley Volcanic Park	Sbly
Skyline serpentine area	Skyln
Springtown	Sprgtwn
Strawberry Canyon, Berkeley	StrwCyn
Sunol Regional Wilderness	Sunol
Sycamore Grove	SycGrv
Tilden Regional Park	Tldn
Wildcat Canyon	WldctCyn

Mount Diablo State Park

This map is not intended to provide exhaustive coverage. Rather it presents an easy-to-use layout of the relationships of major geographical and political features. You can find hiking trails, principal roads, access streets, and major land sites, such as canyons, peaks, and well-known water features. Many obscure or little known topographical, water, or other features have been omitted for the sake of clarity.

The reader will notice that several place names that appear in the Master Plant Location List, Appendix B, are not given. For a more exhaustive treatment of the area, I recommend purchasing Mount Diablo Interpretive Association's fine map of the Mountain.

By studying the major trails and their connectors, you will be assured many hours of pleasurable strolls on the Mountain. These trails take you through all the plant communities found on the Mountain, and introduce you to the majority of trees, shrubs, vines, and wildflowers that occur there. Although these trails are open year round, the novice will find that the best months for observation and identification are February through early June. Fall months are useful for looking at plants in seed and fruit, and for the winter habit of deciduous shrubs and trees.

© Mount Diablo Interpretive Association
Map by Eureka Cartography, Berkeley, CA

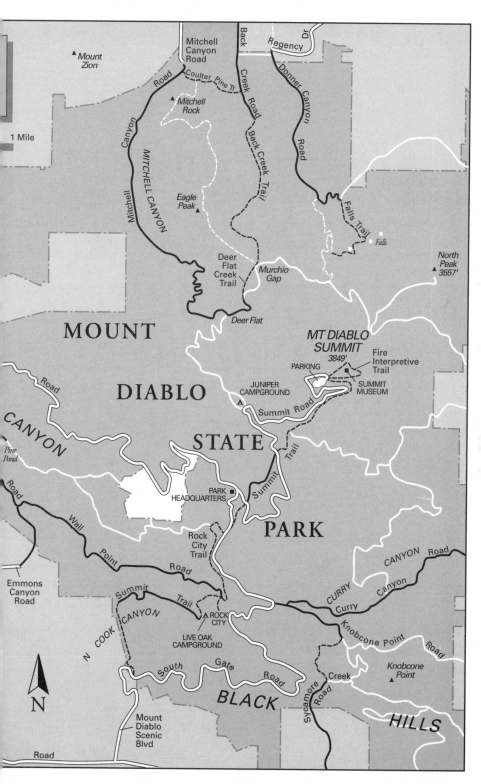

1 Mile

Mount
Zion

Mitchell
Canyon
Road

Back Creek Road

Regency Dr

Coulter Pine Tr

Mitchell
Rock

Canyon

Road

MITCHELL CANYON

Mitchell

Donner Canyon Road

Back Creek Trail

Eagle
Peak

Falls Trail

Falls

North
Peak
3557'

Deer
Flat
Creek
Trail

Murchio
Gap

Deer Flat

MOUNT

MT DIABLO
SUMMIT
3849'

PARKING

Fire
Interpretive
Trail

SUMMIT
MUSEUM

JUNIPER
CAMPGROUND

Summit Road

DIABLO

Road

CANYON

Pine
Pond

STATE

Summit Trail

Road

Wall

PARK

PARK
HEADQUARTERS

Rock
City
Trail

CANYON Road

Point

Road

CURRY

Curry Canyon

Emmons
Canyon
Road

Summit

Trail

ROCK
CITY

Knobcone Point Road

N COOK CANYON

LIVE OAK
CAMPGROUND

South Gate Road

Sycamore

Creek

Road

Knobcone
Point

N

BLACK

Mount
Diablo
Scenic
Blvd

HILLS

Road

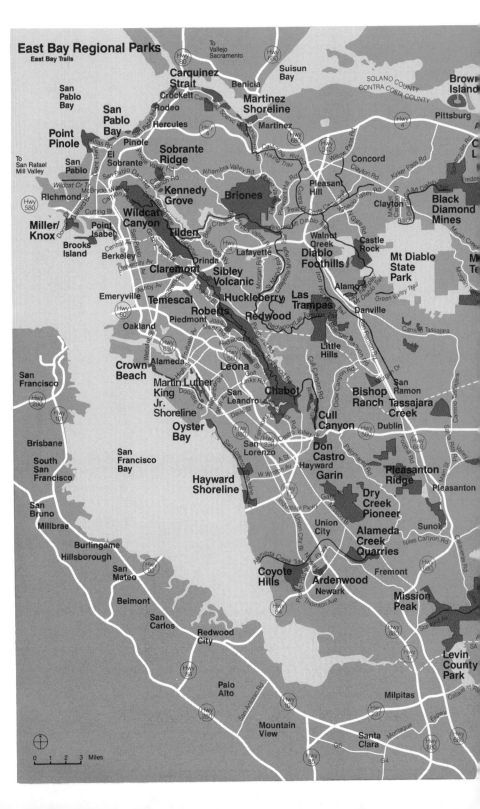

East Bay Regional Parks

This map was made available courtesy of the East Bay Regional Park District. It is intended to give the reader an overview of the locations and positions of the major regional parks with respect to the local geographic features, such as San Francisco Bay, major towns and cities, major arteries, and important roads that interface with the parks themselves. The map has purposely been kept as simple as possible; exhaustive detail would have interfered with overall readability.

If you're contemplating a journey to one of the parks, be sure to consult the directions given under each park heading in Appendix A. Most parks have brochures with more detailed maps that portray trails.

You will notice that most plant entries in Appendix B do not specify the exact location of the plant within the park. This was done to protect sites of rare or unusual plants. (Unfortunately, there is a small minority of people who take advantage of such information.)

I hope you will spend many pleasurable hours pursuing the challenge of finding and identifying the diverse plant life waiting to be encountered in our many parks. These parks provide us with places that remain wild and beautiful.

Appendix B
The Master Plant Location List

This list includes many species that are not described in detail in the Encyclopedias, for lack of space. Still, the list is not exhaustive for our area, which may have as many as 1,500 species and varieties. Most of the plants you're likely to encounter are included.

If you scan the list of symbols for map locations, page 290, the intuitive symbols for parks that are listed below for species will become much clearer. A "?" after a symbol indicates that that particular plant may no longer occur at that location or that a reported citing has not been recently verified.

Important Note: Nonnative, introduced, or alien plants are not included in the following list, since their location may change from year to year.

Trees

Acer macrophyllum (bigleaf maple): **Chbt, MtDbl-MtchCyn-SycCyn-PnCyn, Rdwd, Tldn**

A. negundo (box elder): **Chbt, MtDbl-PnCyn, Niles Cyn, Pinehurst near Rdwd**

Aesculus californica (California buckeye): **many sites with oaks and digger pine**

Alnus rhombifolia (white alder): **MtDbl-MtchCyn-SycCyn, Rdwd, Tldn**

Arbutus menziesii (madrone): **Chbt, MtDbl-PnCyn-SycCyn-DnCkCyn, Rdwd, Tldn**

Corylus cornuta californica (California hazelnut): **Chbt, Hklbr, Rdwd**

Cupressus sargentii (Sargent cypress): **CdrMt**

Juglans hindsii (black walnut): **MnsRd, MtDbl-Mtch-BckCyn-PnCyn, Rdwd**

Juniperus californica (California juniper): **DlVll, MnsRd, MtDbl-BckCyn-JnCmp-EglPt**

Myrica californica (California bayberry): **Chbt, Gtwy, Hklbr, Rdwd, Tldn-JwlLk**

Pinus attenuata (knobcone pine): **Ebmd, FlkrRdg, MtDbl-KnbcnRdg**

P. coulteri (Coulter pine): **BlkDmnd, MtDbl-BckCyn-DnnrCyn-MtchCyn**

P. sabiniana (digger pine): **MtDbl, MnsRd**

Platanus racemosa (western sycamore): **DlVll, Grn, MrshCrk, MrgTrt, MtDbl-SycCyn, PlsntRdg, Sunol**

Populus fremontii (Fremont cottonwood): **BlkDmnd, MrgTrt, MtDbl-MtchCyn-PnCyn**

Quercus agrifolia (coast live oak): **Brns, Chbt, MrgTrt, MtDbl, OkHlls, Tldn**

Q. chrysolepis (goldcup oak): **HklbrTrl, JqnMllr, MrgTrt, MtDbl-smt-SycCyn**

Q. douglasii (blue oak): **Brns, MnsRd, MrgTrt, MtDbl-PnCyn-MtchCyn, PlsntRdg**

Q. garryana (Garry oak): **Brns, Trmps, Fossil Rdg, MtDbl-grdn of Jungle Gods**

Q. kelloggii (California black oak): **MrgTrt, MtDbl-RckCty-PnCyn, Wildcat Cyn Rd**

Q. lobata (valley oak): **Brns, FrklnCyn, MrgTrt, MtDbl-Bck-MtchCyn-PnCyn**

Q. wislezenii (interior live oak): **Chbt, MrgTrt, MtDbl-DnnrCyn-smt**

Salix hindsiana (sandbar willow): **AntDns, MtDbl-MtchCyn**

S. laevigata (red willow): **multiple sites in canyon bottoms**

S. lasiolepis (arroyo willow): **multiple sites in canyon bottoms**

Sambucus racemosa (red elderberry): Tldn-JwlLk, WldctCyn

S. mexicana (blue elderberry): many sites in canyons

Sequoia sempervirens (coast redwood): Cyn, Hklbr, JqnMllr, Rdwd

Umbellularia californica (California bay laurel): many sites, with madrones and live oaks on hillsides, inland on canyon bottoms

Shrubs

Adenostoma fasciculatum (chamise): multiple sites in chaparral

Amelanchier pallida (service berry): Brns, Chbt, MnsRd, Ohln, Sunol, Trmps, Vqrs

Arctostaphylos auriculata (Mt. Diablo manzanita): MtDbl-SycCyn-DnCkCyn-EmnsCyn-KnbcnPt

A. crustacea (brittle-leaf manzanita): Chbt, FlkrRdg, Hklbr

A. glandulosa (Eastwood manzanita): BlkDmnd, MtDbl-KnbcPt-BlkHll-RckCty

A. glauca (big-berry manzanita): MnsRd, MtDbl-DnnrCyn-DrRdg-MtchCyn-JnpCmp

A. manzanita (common manzanita): MrgTrt, MtDbl-DnnrCyn-EmnsCyn-PnCyn

A. pallida (pallid manzanita): ElSbrnt, Hklbr

Artemisia californica (California sagebrush): multiple sites in chaparral

Baccharis pilularis (coyote bush): multiple sites in chaparral or disturbed areas

B. salicifolia (mulefat): MnsRd, MtDbl-BckCyn-MtchCyn-PnCyn

Brickellia californica (California brickel bush): DlVll, MtDbl-BckCyn-MtchCyn, Sunol, SycGrv

Ceanothus cuneatus (buckbrush): multiple sites in chaparral

C. leucodermis (white-thorn): MnsRd

C. sorediatus (Jimbrush): multiple sites in chaparral

C. thyrsiflorus (blueblossom): CrqzStrt, Tldn, WldctCyn

Cercocarpus betuloides (mountain mahogany): Chbt, MnsRd, MrgTrt, MtDbl-smt-DnCkCyn, Trmps

Chrysolepis chrysophylla minor (shrub chinquapin): FlkrRdg, JqnMllr, Hklbr, Sbly

Chrysothamnus nauseosus (rabbitbrush): Dlvll, MtDbl-smt, Ohln, SycGrv

Cornus glabrata (brown dogwood): MtDbl-DnCkCyn-ArryCrr

C. occidentalis (creek dogwood): DmndPrk, Grn, Hklbr, MtDbl-DnCkCyn, Tldn-JwlLk

Dendromecon rigida (bush poppy): BlkDm, BrnsRes, FlkrRdg

Diplacus aurantiacus (bush monkeyflower): multiple sites in chaparral

Dirca occidentalis (western leatherwood): Chbt-BrdTrl, Hklbr, Rdwd-ERgTrl, Tldn

Ericameria arborescens (golden fleece): BlkDmnd, DlVll, MtDbl-EmnsCyn-MrchGp-KnbcnPt

E. linearifolia (goldenbush): MnsRd, MtDbl-MtchCyn-BckCyn-smt

Eriodictyon californicum (yerba santa): Multiple sites on edge of chaparral

Euonymus occidentalis (western burning bush): Tldn near Botanic Gardens

Forestiera neomexicana (desert olive): CrdMt, MnsRd, MrshCrk-VascoRd, MtDbl-MtZion, Vqrs

Fraxinus dipetala (flowering ash): BlkDmnd, MtDbl-DnnrCyn-BckCyn-DrRdg-SlphrSpr, Vqrs

Fremontodendron californicum (fremontia): MrgTrt?, Vqrs

Garrya elliptica (coastal silk tassel bush): Hklbr, Trmps, MtDbl-KnbcnPt-DnCkCyn

G. fremontii (green-leaf silk tassel bush): CdrMt, MtDbl-smt-DrRdg, Trmps?

Gutierrezia californica (California matchstick): AntDns, BlkDmnd, MtDbl-LmRdg, Ohln, Rdwd

Heteromeles arbutifolia (toyon): **multiple sites in chaparral or open woodlands**

Holodiscus discolor (creambush): **multiple sites in forests or dense woodlands**

Keckiella breviflora (gaping penstemon): **MnsRd, SycGrv**

K. corymbosus (red rock penstemon): **MtDbl-smt-JnpCmp-NPk**

Lepechinia calycina (pitcher sage): **BlkDmnd, MtDbl-WlkRdg-smt-DnCkCyn-MtchCyn, Vqrs**

Lepidospartum squamatum (scale broom): **MnsRd, ArryMch**

Lonicera involucrata (twinberry honeysuckle): **Chbt, SPRes, Tldn-JwlLk**

L. subspicata (southern honeysuckle): **MnsRd, MtDbl-NPk**

Lupinus albifrons (blue bush lupine): **MnsRd, MtDbl-DnCkCyn-MtchCyn-DrFlt-smt, Tldn**

Malacothamnus fasciculata (bush mallow): **MnsRd?, MtDbl-LmRdg**

M. fremontii (Fremont bush mallow): **BlkDmnd, MtDbl-BckCyn-EmnsCyn-RckCty-rdg MtchCyn**

Oemleria cerasiformis (oso berry): **Ebmd, Hklbr, MrgTrt, MtDbl-MtchCyn-PnCyn, Rdwd, Tldn**

Physocarpus capitatus (ninebark): **Chbt, Grn, Rdwd, Tldn-JwlLk**

Pickeringia montana (chaparral pea): **FlkrRdg, MrgTrt, MtDbl-DnCk-EmnsCyn-MtchCyn-RckCty, SLRes, Vqrs**

Prunus emarginata (bitter cherry): **Brns, Chbt, MtDbl-PnCyn-SycCyn, Rdwd, Trmps**

P. subcordata (Sierra plum): **MrshCrk, MtDbl-JnpCmp-DrFlt-GrpSpr**

P. virginiana demissa (choke-cherry): **MnsRd, MrgTrt, MtDbl-GrnsdTrl-MtZn-GrpSpr, Ohln, Sunol**

Ptelea crenulata (hopbush): **BlkDmnd, CrqzStrt, MrgTrt, MtDbl-MrchGt-MtchCyn-PnCyn**

Quercus berberidifolia (scrub oak): **many sites in chaparral**

Q. durata (leather oak): **CdrMt, MtDbl-MrchGp-MtchCyn-EglRdg-DnnrCyn**

Rhamnus californica (coffee berry): **multiple sites in woodlands and chaparral**

R. crocea and *ilicifolia* and vars. (redberry buckthorn): **Chbt, Grn, MnsRd, MrgTrt, MtDbl-Mtchl-DnnrCyns, Trmps**

Rhus trilobata (sourberry): **MrgTrt, MtDbl-MtchCyn-DnnrCyn-DrFlt-PnCyn**

Ribes aureum (golden currant): **CdrMt, MnsRd, WldctCyn?**

R. californicum (hillside gooseberry): **multiple sites in brush or forests**

R. malvaceum (chaparral currant): **MtDbl-MtchCyn-BckCyn-EglRdg-smt**

R. menziesii (canyon gooseberry): **multiple sites in woods or forests**

R. quercetorum (oak-leaf gooseberry): **MnsRd, Vqrs**

R. sanguineum glutinosum (flowering currant): **Chbt, Hklbr, Rdwd, Tldn**

R. speciosum (fuchsia-flowered gooseberry): **MssnPk, MtDbl-plntd DnnrCyn**

Rosa californica (California wild rose): **Hklbr, MnsRd, MtDbl-MtchCyn-BckCyn-PnCyn, Tldn**

R. gymnocarpa (woodrose): **multiple sites in dense woods or forests**

Rubus parviflorus (thimbleberry): **Chbt, Hklbr, MtDbl-AldSpr-SlphrSpr, Rdwd, Tldn**

Salvia mellifera (black sage): **BlkDmnd, MtDbl-numerous sites, Trmps**

Senecio douglasii (Douglas's bush senecio): **AntDns, Sunol, SycGrv**

Solanum umbelliferum (blue witch): **Chbt, MnsRd, MtDbl-EmnsCyn-MtchCyn-PnCyn, Tldn**

Symphoricarpos albus (bush snowberry): **multiple sites in woodlands and forests**

S. mollis (creeping snowberry): **multiple sites in woodlands and forests**

Toxicodendron diversilobum (poison oak): **multiple sites in many habitats**

Vaccinium ovatum (evergreen huckleberry): **Chbt, Hklbr, JqnMllr, Rdwd**

Vines

Aristolochia californica (California pipevine): **Brns, CrqzStrt, Trmps, WldctCyn**

Calystegia occidentalis (wild morning glory): **multiple sites**

Clematis lasiantha (virgin's bower): **MnsRd, MtDbl-EmnsCyn-smt**

C. ligusticifolia (virgin's bower): **multiple sites in riparian areas**

Cucurbita foetidissima (coyote gourd): **SycGrv?**

Lathyrus jepsonii (Jepson's sweet pea): **AntDns, BrwIsl**

L. vestitus and vars. (wild sweet pea): **multiple sites in dense woods or forest**

Lonicera hispidula (vine honeysuckle): **multiple sites, often in riparian areas or oak woods**

L. interrupta (vine honeysuckle): **MnsRd?, MtDbl-MtchCyn-WtrflTrl-RhnCyn**

Marah fabaceus (common manroot): **multiple sites**

M. oreganus (Oregon manroot): **Hklbr?, Tldn?**

Rubus ursinus (native blackberry): **multiple sites, particularly canyon bottoms**

Vicia americana (American vetch): **MnsRd, MtDbl-CurCyn-SycCyn**

V. gigantea (giant vetch): **Chbt, OkHlls**

Vitis californica (California wild grape): **Alhambra Vlly Rd, GrzPk, MrgTrt, MtDbl-MtchCyn-PnCyn**

Wildflowers

Achillea millefolium (yarrow): **multiple sites in open places**

Achyrachaena mollis (blow wives): **multiple sites in grasslands**

Actaea rubra arguta (baneberry): **Chbt, GrzPk, Hklbr, Rdwd, Tldn**

Agoseris grandiflora (large-flowered native dandelion): **MtDbl-RckCty-EglRdg-CurrCyn**

A. heterophylla (native dandelion): **MtDbl-BlckbrSpr-PnCyn-SycCyn**

A. retrorsa (native dandelion): **MtDbl-smt**

Alchemilla occidentalis (western lady's mantle): **MtDbl-PnCyn-BlkHlls-SycCyn**

Allium crispum (wavy onion): **CrrlHw, Vqrs**

A. acuminatum (Hooker's onion): **MtDbl-smt**

A. amplectens (white-flowered lava onion): **CdrMt, MrgTrt, Ohln**

A. falcifolium (sickle-leaf onion): **CdrMt, MtDbl-smt-EglPk, Skyln**

A. serra (pink onion): **MrgTrt?, MtDbl-MtchCyn-DrRdg-EglRdg-NPk, Sunol**

A. unifolium (single-leaf onion): **Chbt, MrgTrt-RndVlly, MtDbl-WtrflTrl-NPk-MtchCyn-WhtCyn-DvlsElbw, Ohln**

Allophyllum divaricatum (no common name): **CrrlHw, Ohln**

A. gilioides (no common name): **Ohln, Vqrs**

Amsinckia grandiflora (large-flowered fiddleneck): **Lvrm**

A. menziesii intermedia (common fiddleneck): **multiple sites in grasslands**

Anaphalis margaritacea (pearly everlasting): **JqnMlr, OkHlls-SklnBlvd, Rdwd**

Anemopsis californica (yerba mansa): **JerryIsl, Lvrm, MtDbl-SlphrSpr, Vqrs**

Angelica tomentosa (angelica): **Chbt, Hkbl, MtDbl-MtchCyn, Sbly, Tldn**

Antirrhinum hookerianum (Hooker's snapdragon): **MtDbl-NPk-OystrRdg-MtchCyn**

A. kelloggii (Kellogg's snapdragon): **MtDbl**

A. vexillo-calyculatum (wild snapdragon): **Brns, CdrMt, Chbt, MtDbl-smt-SnbdCrkTrl, PlsntRdg, Sunol**

Apocynum androsaemifolium (dogbane): **Chbt, MtDbl-CurrCvCyn, SLRes, Trmps**

A. cannabinum (Indian hemp): **BrwIsl, MnsRd, MtDbl-PnPnd-DnnrCyn-CurrCyn, Sunol**

Aquilegia formosa (red columbine): **Chbt-ClmbnTrl, Hklbr, MtDbl-MtchCyn-DvlsElbwTrl, Skyln-CrstwdRd**

Arabis breweri (Brewer's rock cress): **MtDbl-smt**

A. glabra (tower mustard): **MtDbl-MtchCyn**

Aralia californica (elk clover): **Grn, Hklbr, MtDbl-DnnrCyn, Rdwd-FrnchTrl**

Arenaria douglasii (Douglas's sandwort): **frequent in open rocky sites**

A. macrophylla (perennial sandwort): **MtDbl-smt**

Argemone munita (prickly poppy): **BlkDmnd, MtDbl-MtchCyn-WhtCyn**

Arnica discoidea (rayless arnica): **Gtwy, Hklbr, MtDbl-smt, Rdwd, Tldn**

Artemisia douglasiana (mugwort): **multiple sites in scrub and riparian places**

A. dracunculus (wild tarragon): **dry stream beds in southern Alameda Co.**

Asarum caudatum (wild ginger): **Hkbl, JqnMllr-PalsClrdoTrl, Rdwd, Sbly, Tldn**

Asclepias californica (California milkweed): **DrCrk, MtDbl-NrthGtRd-WtrflTrl-MtchRck-DrFlt**

A. fascicularis (whorled milkweed): **many sites on roadbanks, openings, springs**

Aster chilensis (wild aster): **Chbt, Rdwd, Tldn**

A. radulinus (broadleaf aster): **many sites in woodlands and forests**

Astragalus asymmetricus (rattlepod): **CrqzStrt, Vqrs**

A. gambellianus (Gambel's locoweed): **many sites in grasslands but may be overlooked**

A. oxyphysus (rattlepod): **BlkDmnd, MnsRd?, Vqrs**

Athysanus pusillus (no common name): **common in grasslands but overlooked**

Baccharis douglasii (Douglas's baccharis): **Brns, delta, MtDbl-RslmnPrk-WllwSpr, Tldn**

Balsamorrhiza deltoidea (balsamroot): **BlkDmnd, CntrLm, FrmntRdg**

B. macrolepis (balsamroot): **ArryMch, Lvrm-GrnvllRd**

Barbarea orthoceras (winter cress): **frequent in temporarily damp sites**

Berula erecta (water parsnip): **Grn, MtDbl-MosesSpr-MplSpr**

Bidens laevis (native tickseed): **Delta**

Blennosperma nanum (glue-seed): **many sites in moist grasslands**

Boisduvalia spp. (no common name): **MtDbl-BckCyn-MosesSpr, Ohln, PlsntRdg, Sunol**

Boschniakia strobilacea (cone flower): **ElSbrnt, Ebmd, FlkrRdg, Sbly**

Bowlesia incana (no common name): **MtDbl-DrRdg-NPk-LmRdg-CurrCvTrl, SycGrv, Trmps, Vqrs**

Brodiaea elegans (harvest brodiaea): **Chbt, MrgTrt, MtDbl-RckCty-MtchCyn-PnCyn, Tldn?**

Calandrinia ciliata (red maids): **multiple sites in grasslands**

Calochortus albus (white fairy lantern): **BlkDmnd, Chbt, Grn, MnsRd, Trmps**

C. invenustus (mariposa tulip): **CdrMt, Vqrs**

C. luteus (yellow mariposa tulip): **Brns, JqmMll?, MnsRd, MtDbl-RckCty-PnCyn-KnbcnPt, Rdwd-Rbrts**

C. pulchellus (Mt. Diablo globe tulip): **MrgTrt, DblFthl, MtDbl**

C. splendens (lilac mariposa tulip): **MtDbl-SthGtRd-lwrSycCyn-WllPtRd**

C. superbus (superb mariposa tulip): **DrCrk, FrmntRdg**

C. umbellatus (Oakland star tulip): **JqnMllr, MtDbl-BlckRgRd-SycCyn-KnbcnPt**

C. venustus (white mariposa tulip): **MnsRd, MtDbl-smt-EglRdg-DnnrCyn-NPk, Tldn**

Calycadenia multiglandulosa (rosinweed): **Skln**

C. truncata (rosinweed): **MtDbl-MtchCrk-DnnrCyn-JnpCmp**

Calyptridium monandrum (no common name): **CrrlHw**

Calystegia malacophylla (wooly morning glory): **MnsRd, MtDbl-MrchGp-MtchCyn-smt**

C. subacaulis (creeping morning glory): **many sites on open, stony soils or along edge of chaparral**

Camissonia boothii decorticans (bottle-scrubber): **Lvrm**

C. contorta (suncups): **Ohln?, Sunol**

C. graciliflora (suncups): **CrrlHw, MnsRd**

C. intermedia (suncups): **BlkDmnd, MtDbl**

C. micrantha and vars. (small-flowered suncups): **AntDns, MtDbl-smt, Trmps**

C. ovata (golden eggs): **Chbt, Rdwd, UCBrkl**

Campanula exigua (rock harebell): **CdrMt, MtDbl-smt-EglRdg-RhnCyn, Ohln**

Castilleja affinis (Indian paintbrush): **MtDbl-MtchCyn-DnCkCyn**

C. applegatei (mountain paintbrush): **MnsRd, MtDbl-smt, Vqrs**

C. foliolosa (wooly paintbrush): **many sites in chaparral**

C. wightii (coastal paintbrush): **Chbt, Rdwd, StrwCyn**

Centaurium davyi (canchalagua): **Brns, PtMlt, SPRes**

C. muehlenbergii (canchalagua): **BrwIsl, Skyln**

Chaenactis glabriuscula (yellow pincushions): **CdrMt, MtDbl-DvlsPlpt**

Chlorogalum pomeridianum (soap plant): **multiple sites in grasslands and open woodlands**

Chorizanthe membranacea (spine flower): **MnsRd, MtDbl-smt-JnpCmp-NPk-SlvSpr, Ohln**

C. polyganoides (spine flower): **Ohln**

Chrysopsis villosa (golden aster): **Brns, MtDbl-PnCyn-MtZn-OystrPt, Sbly, Tldn**

Cicendia quadrangularis (yellow gentian): **MtDbl-FsslRdg-RckCty, Vqrs**

Cicuta douglasii (water hemlock): **Chbt-BrdTrl, Ohln, LkTmscl?**

Cirsium breweri (Brewer's thistle): **MtDbl-MplSpr-WllwSpr**

C. occidentale and vars. (cobweb thistle): **MnsRd, MtDbl-smt-MtchCyn, PlsntRdg, Trmps**

Clarkia breweri (fairy fans): **CdrMt, MnsRd**

C. concinna (red ribbons): **Brns, MrgTrt, MtDbl-smt-NPk, PlsntRdg, Trmps, Vqrs**

C. franciscana (San Francisco clarkia): **Skyln**

C. gracilis (graceful clarkia): **CdrMt, MtDbl, SycGrv**

C. purpurea and vars. (winecup clarkia): **multiple sites in grasslands and woodlands**

C. rubicunda (ruby chalice clarkia)

C. unguiculata (elegant clarkia): **Brns, MnsRd, MtDbl**

Clintonia andrewsiana (bead lily): **Rdwd**

Collinsia bartsiaefolia (Chinese houses): **CdrMt, Ohln, Trmps**

C. concolor (Chinese houses): **Sunol**

C. heterophylla (Chinese houses): **many sites in woodlands**

C. parviflora and *C. sparsiflora* (blue-eyed Mary): **many sites in woodlands**

C. tinctoria (Chinese houses): **MtDbl-smt-EglRdg-RnsmPt**

Collomia heterophylla (no common name): **FlkRdg, Hklbr**

Corallorhiza maculata (spotted coral root): **Brns, Hklbr, Rdwd, Tldn**

Cordylanthus mollis and vars. (pelican beak): **PtPnl, Sprtn**

C. nidularis (pelican beak): **MtDbl-BldRdg**

C. pilosus (pelican beak): **Brns, MtDbl-DrFlt-PnCyn, Vqrs**

Corethrogyne filaginifolia (wooly aster): **Chbt, CrrlHw, FrmntRdg, PlsntRdg, Sunol**

Crassula spp. (pygmy weed): **many sites in temporarily wet, open places**

Cryptantha spp. (popcorn flowers): **multiple sites in grasslands and dry open spots**

Cuscuta spp. (dodder): **widespread parasite, locally abundant**

Cynoglossum grande (hound's tongue): **multiple sites in oak and foothill woodlands**

Datisca glomerata (Durango root): **MtDbl-BckCyn-MtchCyn, Sunol**

Datura wrightii (Jimson weed, angel's trumpet): **LmRdg, MnsRd**

Daucus pusillus (native carrot): **MtDbl-EmnsCyn-RsslmnPrk-BlckHlls-OystrRdg**

Delphinium californicum (California larkspur): **MnsRd, MtDbl-EglRdg-JnpCmp**

D. decorum (blue larkspur): **MtDbl-DrFlt-RslmnPrk-NPk-SycCyn**

D. hesperium (western larkspur): **MnsRd, MtDbl-DnnrCyn-PnCyn-CurrCyn**

D. nudicaule (scarlet larkspur): **MtDbl-BckCyn-smt-MtchCyn-DnCkCyn-NPk**

D. patens (blue larkspur): **many sites in woodlands**

D. variegatum (royal larkspur): **MnsRd, MtDbl-OakKnll-near smt**

Dentaria californica (milkmaids): **widespread in woodlands and along forest edges**

Dicentra chrysantha (golden eardrops): **BlkDmnd, MtDbl-smt-MtchCyn-WhtCyn, Trmps?**

D. formosa (western bleeding heart): **Chbt-RdwdRd, Ebmd-above Cyn**

Dichelostemma capitatum (blue dicks brodiaea): **multiple sites in grasslands and woodlands**

D. congestum (ookow): **Brns, MtDbl-RckCty-DnCkCyn-MtchCyn-PnCyn, Tldn**

Disporum hookeri (Hooker's fairy bells): **AlmoCyn, Chbt, ClrmntCyn, Hklbr, JqnMllr?, Rdwd?, LkTmscl**

Dodecatheon clevelandii patulum (padre shooting star): **BlkDmnd, MnsRd**

D. hendersonii (Henderson's shooting stars): **many sites in open woodlands or grasslands**

Downingia spp. (downingias): **Byrn, Sprgtwn**

Dudleya cymosa (hot rock dudleya): **MnsRd?, MtDbl-smt-BckCyn-MtchCyn-EglPt-NPk**

Echinodorus berteroi (water plantain): **Chbt, SLRes**

Emmenanthe penduliflora (whispering bells): **many sites in rocky spots or after burns, MtDbl-smt-EmnsCyn-OystrRdg**

Epilobium brachycarpum (weedy epilobium): **multiple disturbed or open places**

E. ciliatum watsonii (willow herb): **many temporarily wet sites**

E. minutum (tiny-flowered epilobium): **MtDbl-smt**

Epipactis gigantea (brook orchid): **MtDbl-PnCyn, Tldn-sprgInsprtnPt**

Eremocarpus setigerus (turkey mullein): **widespread along roadsides and disturbed sites**

Erigeron foliosus (leafy daisy): **MtDbl-Rndtp**

E. inornatus (rayless daisy): **MtDbl-smt**

E. philadelphicus (common daisy): **MrgTrt, MtDbl-PnCyn-RslmnPrk-BlckbrSpr, Sunol, Tldn**

Eriogonum gracile (graceful buckwheat): **MtDbl-EglPt-NPk-smt-TrtlRck**

E. inerme (buckwheat): **MtDbl-NPk-smt**

E. nudum (naked stem buckwheat): **multiple sites on dry, rocky slopes**

E. truncatum (buckwheat): **MtDbl-MtZn-AlmoCyn**

E. umbellatum (sulfur buckwheat): **CdrMt, MtDbl-smt-NPk-DvlsPlpt, Ohln**

E. vimineum (pink-flowered buckwheat): **MnsRd?, Rdwd**

E. wrightii (Wright's buckwheat): **CrrlHw, MnsRd**

Eriophyllum confertiflorum (golden yarrow): **Brns, CrqzStrt, MnsRd, MtDbl-smt-DnCkCyn, Rdwd**

E. lanatum (Oregon sunshine): **MrgTrt, MtDbl-many sites**

Eryngium aristulatum (button parsley): **MtDbl-EmnsCyn, PtPnl, Sprtn, Vqrs**

E. armatum (button parsley): **BrnsRes, MrshCrk**

E. articulatum (button parsley): **AntDns, BrwIsl**

Erysimum capitatum (foothill wallflower): **MnsRd, MtDbl-DnCkCyn-smt-MtchCyn, Sunol**

Eschscholzia caespitosa (foothill poppy): **CdrMt, MnsRd?, SycGrv, Vqrs**

E. californica (California poppy): **multiple sites in grasslands**

Euphorbia crenulata (native spurge): **MnsRd, MtDbl-MtchCyn-RslmnPrk-SycCyn, Vqrs**

Fragaria vesca (woodland strawberry): **widespread in dense woodlands and forests**

Fritillaria agrestis (stink bells): **CrrlHw, Lvrm**

F. falcata (sickle-leaf fritillaria): **CdrMt, MnsRd**

F. lanceolata (mission bells): **Chbt, JqnMllr, MrgTrt, Tldn**

F. liliacea (white fritillary): **Brns-MtDblTrl, DblFthl**

Galium spp. (bedstraw): **multiple sites in disturbed places, chaparral, and woodlands**

Geranium molle (wild geranium): **many sites in open woods**

Gilia achillaefolia multicaulis (gilia): **many sites in open places, including grasslands**

G. capitata (globe gilia): **AntDns, CrrlHw, MtDbl-smt-NPk, Sunol**

G. tricolor (Bird's eye gilia): **MrgTrt, MtDbl-DrFlt-WtrflTrl-DnnrCyn-WndyPt**

Githopsis specularioides (looking glass plant): **MtDbl-smt-MtchCyn-NPk-SycCyn, Ohln**

Glycyrrhiza lepidota (wild licorice): **CdrMt, MtDbl-PnCyn, Sunol**

Gnaphalium spp. (cudweed): **multiple sites in open, often disturbed places**

Grindelia spp. (gumweed): **multiple sites in open chaparral and scrub**

Helenium puberulum (rosilla): **Chbt, Grn, MtDbl-SycCyn-PnCyn-MtchCyn, Tldn?**

Helianthella castanea (Mt. Diablo sunflower): **multiple sites around Mt. Diablo**

Helianthemum scoparium (rush-rose): **BlkDmnd, MrgTrt, MtDbl-EmnsCyn-RckCty-LmRdg**

Helianthus californicus (California sunflower): **BrwIsl, DlVll, MtDbl-MtchCyn-DnnrCyn, Sunol**

H. gracilentus (serpentine sunflower): **MtDbl-DnnrCyn-MtchCyn-BlkHlls**

Hemizonia congesta and vars. (tarweed): **multiple open sites**

H. fitchii (spiny tarweed): **many sites in disturbed grassland**

Heracleum lanatum (cow parsnip): **Hklbr, MtDbl-MtchCyn-Spr-SycCyn, Tldn**

Hesperolinon breweri (Brewer's flax): **MrgTrt, MtDbl-HthrngtnLp-MtchCyn-WhtCyn-MrchGp, Vqrs**

H. californicum (California flax): **CrrlHw, MtDbl-BlckHll-WllPtRdg, FrmntRdg**

Heterocodon rariflorum (no common name): **by Mtchl Crk?**

Heuchera micrantha (common alumroot): **Hklbr, MtDbl-DnCkTrl-DnnrCyn-SycCyn-FsslRdg, NilesCyn, Rdwd**

Hieracium albiflorum (white hawkweed): **Brns, Chbt, Cyn, FlkrRdg**

Holocarpha heermanii (Heerman's tarweed): **abundant in grasslands**

Horkelia spp. (no common name): **AlbHll, ClrmtCyn, CrqzStrt, Ebmd-nearChbt, MrgTrt, MtDbl-RslmnPrk-MplSpr**

Hydrocotyle spp. (pennywort): **DowWetlndsPres**

Hypericum formosum (native St. Johnswort): **Chbt, MtDbl-SlvrSpr-CynSpr-EldrSpr**

Iris douglasiana (Douglas iris): **ClrmtCyn, JqnMllr, Hklbr, Rdwd, Tldn**

I. longipetala (coast iris): **Chbt?, WldctCyn**

Isopyrum occidentale (western rue-anemone): **Ohln-IndnCrk**

I. stipitatum (rue-anemone): **CdrMt, MnsRd**

Lasthenia spp. (goldfields): **multiple sites in grasslands**

Layia fremontii (Fremont's tidy-tips): **Vqrs**

L. gaillardioides (yellow tidy-tips): **MrgTrt, MtDbl-MtchCyn-EglRdg, Ohln, Sunol**

L. glandulosa (desert tidy-tips): **Vqrs**

L. hieracioides (tidy-tips): **MtDbl-DrFlt-BlkbrSpr**

L. platyglossa (common tidy-tips): **many sites in grasslands**

Lepidium nitidum (pepper-grass): **widespread in grassy places**

Lessingia germanorum (no common name): **AntDns, MtDbl-LmRdg**

Lewisia rediviva (bitterroot): MrgTrt, MtDbl-smt-RnsmPtTrl, Ohln

Lilium pardalinum (leopard lily): Chbt-BrdTrl, Gtwy, MtDbl-PnCyn?, Tldn-sprg near InsprtnPt

Limnanthes douglasii and vars. (meadow foam): Byrn, Vqrs

Linanthus androsaceus (linanthus): MtDbl-DrFlt-EglRdg-ArryCrr

L. bicolor (linanthus): multiple sites in grasslands or open woodlands

L. ciliatus (mustang clover): many sites in open woodlands or grasslands

L. dichotomus (evening snow): CdrMt?, CrrlHw, MtDbl-HthrgtnLp

L. grandiflorus (large-flowered linanthus): MnsRd?

L. montanus (mountain linanthus): MtDbl-WtrflTrl?-MtchCyn?

L. pygmaeus (pygmy linanthus): MtDbl-OystrRdg-SycCyn

Linaria canadensis (toadflax): Byrn, Chbt, CrqzStrt, MtDbl-DnCkCyn-EglRdg, Vqrs

Lithophragma affinis (woodland star): many sites in woodlands and forest edges

L. heterophyllum (woodland star): many sites in woodlands and forest edges

Lomatium californicum (California biscuit root): Grn, MtDbl-PrsptrGp-MtchCyn?, Sunol-LttlYsmt

L. caruifolium (biscuit root): AlbHll, CdrMt, MrgTrt?

L. dasycarpum (biscuit root): widespread in rocky open places

L. macrocarpum (biscuit root): common in grassy places

L. nudicaule (swollen-stem biscuit root): MtDbl-NPkTrl-DrFlt, Ohln

L. utriculatum (biscuit root): widespread in grasslands

Lotus crassifolius (thick-leaf lotus): MnsRd, MtDbl-smt

L. humistratus (colchita): multiple sites in open situations

L. micranthus (small-flowered lotus): multiple sites in open situations

L. scoparius (deer broom lotus): many sites on chaparral, scrub, or recently burned areas

L. subpinnatus (no common name): many sites on rocky or bare banks

Lupinus bicolor (dove lupine): multiple sites in grasslands

L. densiflorus (whorled lupine): multiple sites in grasslands or other open situations

L. formosus (woodland lupine): Grn, MtDbl-SycCrk, OkHlls, Tldn, Trmps

L. nanus (sky lupine): frequent in grasslands

L. subvexus (rose lupine): MnsRd

Lythrum californicum (California loosestrife): MtDbl-RslmnPrkRd-MosesRck

Madia elegans (elegant tarweed): many sites in grasslands

M. gracilis (graceful tarweed): MtDbl-EglRdg-OystrRdg

M. madioides (woodland tarweed): Cyn, FlkrRdg, Hklbr, MrgTrt, Sunol?

M. sativa (common tarweed): multiple sites in disturbed places

Malacothrix clevelandii (Cleveland's dandelion): MtDbl-MtchCyn-EglRdg-NPk

M. coulteri (Coulter's dandelion): CrrlHw

M. floccifera (flocked dandelion): Brns, CdrMt, MtDbl-BlkbrSpr-AldrCyn

Meconella oregona (no common name): Sbly

Mentzelia affinis (blazing star): BlkDmnd, CntrLm, Sunol

M. dispersa (blazing star): MtDbl-NPk-smt

M. laevicaulis (common blazing star): DlVll-MnsRd?

M. lindleyi (Lindley's blazing star): NilesCynRd, Ohln, Sunol

M. micrantha (tiny-flowered blazing star): MtDbl-smt

Microseris spp. (native dandelions): many sites in grasslands

Mimulus cardinalis (scarlet monkeyflower): Chbt-BrdTrl, MtDbl-BckCyn-PnCyn-MtchCyn, WldctCyn

M. constrictus (monkeyflower): **MtDbl**

M. douglasii (mickey mouse monkeyflower):
**MnsRd, MtDbl-DnnrCyn-BlkbrSpr,
Rdwd, Skyln**

M. guttatus (golden monkeyflower): **multiple
sites in wet places**

M. kelloggii (Kellogg's monkeyflower):
MnsRd, Vqrs

M. latidens (monkeyflower): **CrrlHw, Sunol**

M. rattanii (monkeyflower): **MnsRd, MtDbl-
JnpCmp-MtchCrk**

M. tricolor (vernal pool monkeyflower): **Vqrs**

M. villosus (silky hair monkeyflower): **CdrMt,
Sunol**

Monardella douglasii (Douglas's coyote mint):
MtDbl-MtchCyn-PnCyn-MrchGp-DrFlt

M. villosa (common coyote mint): **many open
sites as along chaparral or in rocky places**

Monolopia major (monolopia): **CntrLm,
CrrlHw, MtDbl-OystrRdg-RggsCyn**

Montia gysophiloides (spring beauty): **many
sites on rocky or gravelly slopes**

M. perfoliata (miner's lettuce): **multiple sites
in woodlands**

M. spathulata (no common name): **many sites
in open woods**

Muilla maritima (no common name): **Byrn,
FrmntRdg, Sprgtwn, Vqrs**

Myosurus spp. (mousetail): **Byrn, MssnPk,
Sprtwn, Vqrs**

Navarretia mellita (no common name): **many
sites in open, sometimes disturbed places**

N. pubescens (no common name): **many sites
in open places**

N. squarrosus (skunk weed): **many sites in
open, sometimes disturbed places**

Nicotiana bigelovii (native tobacco):
**BlkDmnd?, MnsRd, MtDbl-MtZn,
s.AlamedaCo**

Oenanthe sarmentosa (water parsley): **multiple
sites in marshes and sluggish streams**

Oenothera deltoides cognata (birdcage evening
primrose): **CrrlHw**

Oe. d. howellii (Antioch dunes evening
primrose): **AntDns**

Oe. elata hookeri (Hooker's evening primrose):
AntDns, BrwIsl, MtDbl-HntSpr-SlvrSpr

Orobanche bulbosa (bulbous broomrape):
MtDbl-smt area

O. fasciculata (clustered broomrape): **MtDbl-
MtchCyn-smt**

O. uniflora (single-flowered broomrape):
**MnsRd, MtDbl-MtchCyn-BckCyn-
WtrflTrl-DvlsElbTrl**

Orthocarpus attenuatus (no common name):
many places in grasslands

O. erianthus (cream sacs): **common in grassy
places**

O. lithospermoides (cream sacs): **BlkDmnd,
MtDbl-BlkHlls-EglRdg, Trmps**

O. purpurascens (owl's clover): **multiple sites
in grasslands**

O. pusillus (no common name): **frequent in
grasslands (often overlooked)**

Osmorhiza brachypoda (sweet cicely): **MrgTrt**

O. chilensis (common sweet cicely): **multiple
sites in forests and woodlands**

Oxalis oregana (redwood sorrel): **DmndPrk**

Papaver californicum (fire or flame poppy):
**MtDbl-BckCyn-MtchCyn-EglPk, Trmps,
Vqrs**

Parvisedum pentandrum (annual stonecrop):
MrgTrt, Vqrs

Pectocarya penicillata (comb-fruit): **AntDns,
Lvrm, Sunol**

Pedicularis densiflorus (Indian warrior):
**BlkDmnd, CdrMt, ElSbrnt, MrgTrt,
MtDbl-MrchGt-smt-DnCkCyn-MtchCyn**

Penstemon centranthifolius (scarlet bugler):
BlkDmnd, MtDbl-LmRdg

P. heterophyllus (foothill penstemon): **CdrMt,
MnsRd, MtDbl-smt-DrRdg-MtchCrk,
ShlRdg, Sunol**

Perideridia gairdneri (yampah): **MtDbl-
EmnsCyn?-PnCyn?**

P. kelloggii (Kellogg's yampah): **many places in
rocky meadows, MtDbl-DnCkCyn-PnCyn**

Phacelia breweri (Brewer's phacelia): **Lvrm,
MnsRd, MtDbl-smt-MtchCyn-EglPt,
Ohln**

P. californica (California or rock phacelia): MnsRd?, MtDbl-smt-MtchCyn-EglRdg-NPk, Rdwd, Tldn

P. distans (no common name): common in open places or woodlands

P. nemoralis (bristly phacelia): MtDbl-smt-EglPk, PlstnRdg

P. phacelioides (no common name): MtDbl-smt-EglPt-JmWhtTrl-DrFlt

P. tanacetifolia (tansy-leaf phacelia or caterpillar flower): CntrLm, MtDbl-MtchCyn-DrRdg

Phlox gracilis (annual phlox): multiple sites in open woodlands or grassy places

Pholistoma auritum (fiesta flower): Chbt, Grn, MrshCrk, MtDbl-SycCyn-AlmoCyn, Sunol

P. membranaceum (white fiesta flower): CrrlHw, DlVll, Sunol

Piperia spp. (rein orchid): widespread in dense woodlands and forests but seldom abundant

Plagiobothrys nothofulvus (nievitas): multiple sites in open woodlands or grasslands

P. spp. (popcorn flowers): multiple sites in open, wettish spots

Plantago erecta (annual plantain): multiple sites in grasslands

Platystemon californicus (cream cups): multiple places in grasslands

Plectritis spp. (native valerian): many sites in open woodlands

Pogogyne serpylloides (vernal pool mint): many sites in temporarily wet grasslands

Polemonium carneum (foothill Jacob's ladder): StnbrkCyn, Dublin & Niles Cyn

Polygala californica (California milkwort): Ebmd-above Cyn, FlkrRdg, Rdwd

Polygonum amphibium (water knotweed): BrwIsl, Chbt, Grn, MtDbl-PnPnd, SPRes

Potentilla glandulosa (sticky cinquefoil): frequent in open brushy or wooded places

Psoralea californica (no common name): CdrMt, MtDbl-smt

P. macrostachya (Indian hemp): CdrMt, AntDns

P. physodes (California tea): Chbt, CrqzStrt, Ebmd, Hklbr, MtDbl-DnCkCyn-MtchCyn-PnCyn, Tldn, Trmps

Pterostegia drymarioides (no common name): MnsRd, MtDbl-EmnsCyn-smt

Rafinesquia californica (white chicory): MtDbl-DnCkCyn-EmnsCyn-MtchCyn-JnpCmp-CurrCyn

Ranunculus aquatilis (water buttercup): MtDbl-HddnPnd-TrtlRck, SycGrv

R. californicus (California buttercup): multiple sites on grasslands and open woodlands

Rorripa curvisiliqua (yellow cress): MtDbl-MplSpr

Rumex maritimus (maritime dock): Chbt

R. salicifolius (willow-leafed dock): Chbt, MrgTrt, Sunol, Vqrs

Sagina occidentalis (western pearlwort): frequent in open places, often overlooked

Sagittaria latifolia (arrowhead): Chbt, SLRes

Salvia carduacea (thistle sage): CrrlHw?

S. columbariae (chia): MtDbl-smt-MtchCyn-EglPt-NPk, Ohln, PlsntRdg, Sbly, Sunol

S. spathacea (hummingbird sage): MtDbl-PnCyn-JnpCmp

Sanicula arctopoides (footsteps-to-spring): Rdwd-ERdgTrl

S. bipinnata (poison sanicle): multiple places in open woodlands

S. bipinnatifida (purple sanicle): many places in open woodlands and grasslands

S. crassicaulis (woodland sanicle): multiple places in dense woodlands and forests

S. laciniata (fringe-leaf sanicle): Chbt

S. saxatilis (rock sanicle): MtDbl-smt

S. tuberosa (tuberous sanicle): MnsRd, MrgTrt, MtDbl-MtchCyn

Satureja douglasii (yerba buena): many sites along interface of scrub and woodland or forest

Saxifraga californica (California saxifrage): MnsRd, MrgTrt, MtDbl-MtchCyn-DnnrCyn-CvPt-SycCyn

Scrophularia californica (California bee plant): multiple sites in brushy places and open woodlands

Scutellaria californica (California skullcap): Ebmd, Gtwy, Tldn

S. tuberosa (tuberous skullcap): BlkDmnd, MtDbl-smt-JnpCmp-BckCyn-WllRdg, Ohln, Sunol, SycGrv, Trmps

Sedum spathulifolium (common stonecrop): MtDbl-WtrflTrl-TrtlRck, Sunol-LttlYosmt

Senecio aronicoides (wooly butterwort): MrgTrt, MtDbl-smt-NPk-KnbcRdg

S. breweri (Brewer's butterwort): CrrlHw, MnsRd, MtDbl-MtchCyn-FsslRdg, Ohln, Sunol, Vqrs

Sidalcea malvaeflora (checker bloom): FrklnCyn, PtMlt, Tldn, Vqrs

Silene californica (Indian pink): Chbt?, Cyn, FlkrRdg, Gtwy, MrgTrt, MtDbl-EglRdg-DnnrCyn, Sunol

Sisyrinchium bellum (blue-eyed grass): many sites in grasslands or open woodlands

S. californicum (yellow-eyed grass): Ebmd-near InsprtnPt Tldn, Gtwy

Smilacina racemosa (fat false Solomon's seal): AlamoCyn, MtDbl-KnbcnPt

S. stellata (starry false Solomon's seal): Ebmd, Hklbr, JqnMllr, MtDbl-DnnrCyn-GrpSpr, Tldn

Solidago californica (California goldenrod): many sites in open places

Spergularia macrotheca (sand spurrey): Byrn, Sprtn, Vqrs

S. rubra (sand spurrey): many open sites in disturbed or hardpacked soils

Stachys albens (white-flowered wood mint): AntDns, BrwIsl, SycGrv, Vqrs

S. pycnantha (rock wood mint): MtDbl-MtchCyn-PnCynCastle-DnnrCyn

S. ajugoides rigida (common wood mint): multiple places in woodlands and forests

Stephanomeria virgata (wand-lettuce): MtDbl-EmnsCyn-SlvrSpr, Rdwd

Streptanthus albidus (white jewel flower): MtDbl-MrchGp-EglRdg

S. glandulosus (common jewel flower): MnsRd?, MtDbl-DnnrCyn, Ohln, PlsntRdg

S. hispidus (Mt. Diablo jewel flower): MtDbl-JnpCmp-smt-BckCyn?-WtrflTrl-NPk

S. lilacinus (lilac jewel flower): CrrlHwl, MtDbl-ArryCrr

Stylomecon heterophylla (wind poppy): Brns, DlVll, MrshCrk, MtDbl-DnCkCyn-BckCyn-MtchCyn-smt, Trmps

Tauschia hartweggii (no common name): Grn?, MtDbl-smt-DrFlt-NPk, Tldn

Tellima grandiflora (fringe-cups): Hklbr, JqnMllr, LkTmscl, Rdwd

Thalictrum fendleri polycarpum (foothill meadow rue): Chbt, MrgTrt, MtDbl-BckCyn-MtchCyn-JnpCmp, PlsntRdg, Tldn

Thelypodium lasiophyllum (common thelypodium): widespread in open places

Thysanocarpus spp. (lace- and fringepods): widespread in open, grassy places

Tonella tenella (no common name): MtDbl-DvlsElbw to PrsptrGp, Ohln, Sunol

Trichostema lanceolatum (vinegar weed): many sites in open places or thin grasslands

Trientalis latifolia (star flower): Hklbr, Rdwd, Tldn

Trifolium fucatum (cow clover): MtDbl-MtchCrk, Vqrs

Trifolium spp. (many clover species included here; too numerous to list separately): multiple sites in grasslands, woodlands, or edge of wet places

Trillium chloropetalum (giant trillium): Brns, Chbt-BrdTrl-CscdTrl, MtDbl-SycCyn-GrpSpr, Rdwd

T. ovatum (coast trillium): Cyn, Ebmd-nearTldn, Hklbr, JqnMllr, StrwCyn

Triodanis biflora (no common name): MtDbl-BlkbrSprg?

Triteleia hyacinthina (white brodiaea): FrklnCyn, MrgTrt, MtDbl-DnCkTrl

T. laxa (Ithuriel's spear): **many sites in grasslands and woodlands**

Urtica dioica (stinging nettle): **multiple sites in canyon bottoms along streams**

Verbena lasiostachys (vervain): **many sites in rocky creeks and other open places**

Veronica americana (speedwell): **MtDbl-BlckSpr-SlvrSpr-MplSpr, PlsntRdg, Tldn**

Viola adunca (dog violet): **Hklbr, Rdwd**

V. douglasii (Douglas's violet): **MnsRd**

V. glabella (smooth yellow violet): **Cyn, FlkrRdg, Rdwd**

V. pedunculata (wild pansy): **MnsRd, MrgTrt, MtDbl-many sites**

V. purpurea (oak or pine violet): **MnsRd, MtDbl-smt-MtchCyn-OystrRdg, Ohln, Sunol**

V. sempervirens (redwood violet): **Rdwd**

V. sheltonii (Shelton's violet): **MnsRd, MtDbl-smt, Ohln**

Wyethia angustifolia (narrow-leaf mule's ear): **multiple sites in grasslands or open woodlands**

W. glabra (common mule's ear): **multiple sites in open woodlands**

W. helenioides (mule's ear): **MrgTrt, MtDbl-MtchCyn-smt-CvPt-WndPt**

Zauschneria californica (hummingbird fuchsia): **multiple sites on rocky banks or streamsides**

Zigadenus fremontii (Fremont star lily): **many sites in open woodlands**

Z. paniculatus (no common name): **Vqrs?**

Z. venenosus (death camass): **CdrMt**

Illustrated Glossary

achene. A single-seeded fruit. The fruit is shed as a unit and resembles a seed. A good example is an unshelled sunflower "seed."

alternate. Describes leaf placement in which each leaf is attached at a different level on a stem.

anther. The name for the pollen sacs at the end of each stamen.

appendage. A general term for any outgrowth on some part of a plant. For example, the appendages of daffodil petals produce a trumpet- or cup-shaped structure.

axillary. A bud, flower, or branch that occurs in the angle (axil) between leaf node and stem.

basal. Describes the situation in which leaves occur at the base of the plant, next to the soil.

berry. A fleshy fruit that contains several seeds. Examples include tomatoes, bananas, and eggplants.

bract. Any modified leaf associated with a flower. Bracts may be green, like smaller versions of leaves, or they may be colorful and replace petals (as with the garden calla lily or the Christmas poinsettia).

bristle. A narrow, firm, spinelike or hairlike structure.

bulb. Ball- or globe-shaped underground organ that stores food and water during the dormant period. Consists of fleshy leaf bases surrounding a bud for next year's flowers.

capsule. A dry seed pod that splits open into two or more sections to shed its seeds.

column. The complex structure in orchid flowers that consists of stamen(s) fused to the style and stigma.

achene

alternate

anther

berry

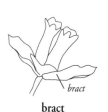

bract

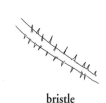

bristle

bulb

capsule

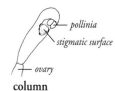

column

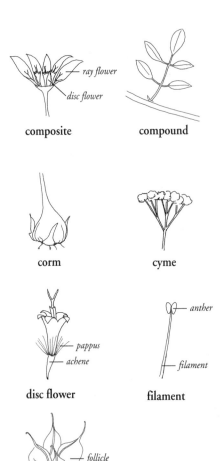

composite

compound

corm

cyme

disc flower

filament

follicle

community. An association of plants growing together, which has recognizable consistency. Most plant communities are defined by their overall appearance and the most obvious or common plants in them (the dominants). Examples include grasslands, oak woodlands, and freshwater marshes.

composite. Describes the grouping together of several flowers to resemble a single large flower. Applied especially to the daisy or sunflower family Asteraceae.

compound. Describes a leaf composed of several distinct parts or leaflets. A bud sits at the base of the leaf, but not at the base of leaflets.

corm. An underground organ similar to a bulb but solid inside (modified starch-storing stem rather than leaf bases).

cyme. Complex arrangement of flowers in umbrellalike, flat-topped clusters.

disc (flower). Describes the tiny, star-shaped flowers in the middle of a composite or daisy head. Each flower has five symmetrical petals, five fused stamens, and two styles.

entire. Describes leaves whose margins are smooth and even, not lobed, scalloped, or toothed.

escape. A nonnative cultivated plant that manages to grow on its own away from gardens.

filament. The stalk or stemlike part of the male structure (stamen) of the flower.

follicle. A seed pod with a single chamber, which opens by one lengthwise slit.

fruit. General term used to describe the mature or ripe ovary of flowering plants. Fruits may split open to release seeds or be fleshy or woody. In the latter case, the whole fruit is shed as a unit.

ground cover. Describes plants that grow low to the ground and spread or propagate themselves horizontally to cover large areas.

gynostegium. The complex reproductive structures of milkweeds in which the stigmas, styles, and stamens are fused together.

half-inferior. Describes a condition where the ovary of the flower is situated below all of the flower parts except for the sepal cup, in which the ovary sits. Typical of flowers in the saxifrage and some members of the rose family.

head. A tight cluster of flowers at the end of a single stem or flowering stalk. Daisies and their relatives have heads of flowers.

herb. Any nonwoody plant. Plants without bark and wood.

inferior. Describes the position of the ovary of a flower below all the other flower parts. Typical of garden fuchsias, currants, and gooseberries.

involucre. A cup-shaped or bowl-shaped row of bracts surrounding a group of flowers.

irregular (flower). Describes flowers whose petals aren't all the same size or shape.

lanceolate. Lance shaped.

legume. A member of the pea family Fabaceae and also their kind of fruit. A legume is a single-chambered seed pod that opens lengthwise by two slits (as with a pea pod). Compare to follicle.

linear. Long, narrow shape, with parallel sides.

lip. The single enlarged lower petal of orchid flowers or one of two sets of petals at the entrance to a flower tube in two-lipped flowers, as in the snapdragon family, Scrophulariaceae, and the mint family, Lamiaceae.

lobed. Describes leaves or petals that have deep, smooth indentations.

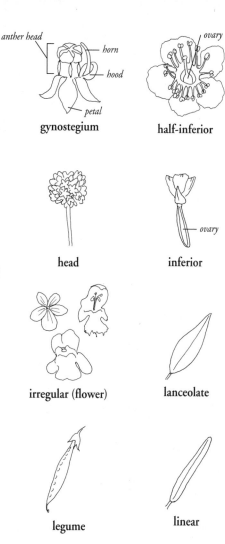

anther head
horn
hood
petal
gynostegium

ovary
half-inferior

head

ovary
inferior

irregular (flower)

lanceolate

legume

linear

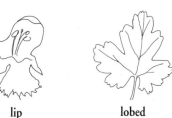

lip

lobed

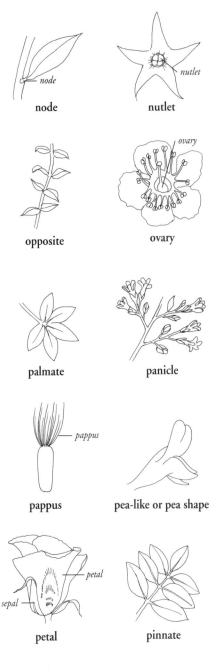

node

nutlet

opposite

ovary

palmate

panicle

pappus

pea-like or pea shape

petal

pinnate

node. The place at which a leaf joins the stem.

numerous. In botany, describes more than twelve of some flower part, such as stamens in buttercup flowers.

nutlet. Small, single-seeded fruits that don't open. Similar to achenes, but the ovary wall is thick and hard. The fruit type typical of the mint family, Lamiaceae, and the borage family, Boraginaceae.

opposite. Describes leaves that are paired on stems.

ovary. The swollen bottom part of the flower's pistil in which the seeds are borne.

palmate. Describes veins, leaf lobes, or leaflets arranged in a fanwise manner, like the fingers on a human hand.

panicle. Complex grouping of flowers consisting of a raceme *of* racemes. See raceme.

pappus. The highly modified sepals of each *individual* composite flower. The pappus usually consists of a row of hairs, bristles, or scales, such as the down on dandelion seeds.

parasite. Plant that depends on other plants for food or water. Many parasites lack green chlorophyll, and so cannot photosynthesize.

pealike or pea shape. Describes a flower in which the petals are arranged in the following manner: the upper as a back banner, the two side petals as wings, and the two fused middle petals between the wings—fashioned into a boatlike structure—as the keel.

perennial. A plant living three or more years and generally not woody.

perfoliate. Describes two leaves that are joined around a stem.

petal. The second whorl or row of parts in a flower; usually brightly colored to attract pollinators.

pinnate. Describes veins, leaf lobes, or leaflets

pistil. The center female part of the flower. Consists of ovary, style, and stigma.

pollination. The process of transporting pollen from the anther of the stamen to the stigma of the pistil. This process is usually carried out by insects or birds.

pollinator. The animal responsible for carrying pollen from the stamen to the stigma.

raceme. Describes flowers arranged along a long stem, each flower borne on a side branch along that stem.

ray flower. The large, showy outer flowers of composites or daisies. The five petals are completely fused to form a tongue- or strap-shaped structure that superficially resembles a single petal.

regular (flower). Describes flowers whose petals are all shaped and arranged alike.

rhizome. Creeping underground stem that is thickened to store food.

rootstock. Underground stem that may probe deeply into the soil or under rocks. Roots develop along this stem.

rosette. A symmetrical set of leaves borne at ground level.

scale. Describes structures (such as the pappus of composite flowers) shaped like fish scales.

schizocarp. An ovary (fruit) that splits into two or more single-seeded segments. Examples include the cheese-wheel-like divisions of mallow fruits and the two sectioned fruits in the parsley family, Apiaceae.

sepal. The outermost whorl of the flower, usually covering and protecting the flower bud.

sheath(ing). Describes the condition in which the base of a leaf forms a partial or complete cylinder around the stem.

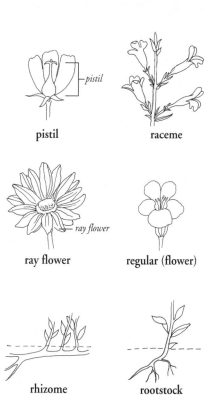

pistil

raceme

ray flower

regular (flower)

rhizome

rootstock

rosette

schizocarp

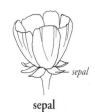

sepal

sheath(ing)

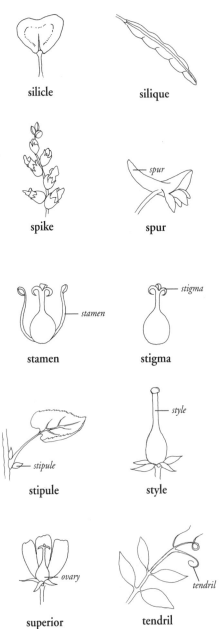

silicle

silique

spike

spur

stamen

stigma

stipule

style

superior

tendril

shrub. A woody plant with multiple stems, not one main trunk.

silicle. Special fruit type in the mustard family, Brassicaceae, in which the ovary (fruit) is as broad as long. Money plant or honesty is a garden flower with conspicuous silicles.

silique. Special fruit type in the mustard family, Brassicaceae, in which the ovary (fruit) is much longer than broad. Siliques are typical of mustards and wallflowers.

simple. Describes a leaf that is not divided into two or more leaflets. Simple leaves may be deeply lobed or slashed, however.

spike. Flower arrangement in which flowers are borne directly along (up) a stem.

spur. The narrow, tapered, pointed nectar sac on such flowers as garden nasturtiums.

stamen. The male part of the flower, located just outside the pistil. Each stamen consists of a stalk (filament) and pollen sacs (anthers).

sterile. Describes flowers that don't make functional stamens or pistils.

stigma. The top part of the pistil, usually enlarged and often sticky or bumpy to accept pollen.

stipule. Pairs of (usually) small, leaflike structures at the base of leaves. Many plants lack stipules altogether.

style. The stalk portion of the pistil extending up from the top of the ovary.

succulent. Any plant with leaves or stems modified to store water.

superior. Describes ovaries that are situated above the attachment of other flower parts.

tendril. A coiled stalk that clasps onto other plants to allow vines to climb.

tree. Woody plants with one or few main trunks and much smaller side branches.

two-lipped. Describes a flower in which sets of petals are situated above and below the opening to the flower tube (throat). Typical of flowers in the snapdragon family, Scrophulariaceae, and mint family, Lamiaceae.

umbel. A flower arrangement in which flowers are borne at the ends of radiating spokelike stalks, much like the ribs of an umbrella.

vein. The vascular strands that carry water and food in leaves. Vein pattern may be important in differentiating major plant groups.

whorl(ed). Leaf arrangement in which three or more leaves are attached to the same level of a stem. Lilies often have whorled leaves.

two-lipped

umbel

vein

vein

whorl(ed)

Index

Note: Page numbers are given in the following order: in the keys; in the encyclopedia descriptions; and in appendix B. *Italicized* numbers indicate encyclopedia descriptions; **bolded** numbers indicate illustration pages; ***bolded and italicized*** numbers indicate a color photograph in the Photo Gallery. Numbers for plant families indicate encyclopedia pages.

Biographies

Glenn Keator grew up in Alameda, California, where, early on, he fell in love with his grandmother's garden. It has been a love affair with plants ever since. He began serious study of wildflowers in junior high, inspired by family camping trips to the Sierra and deserts.

Glenn studied botany at the University of California at Santa Barbara, then did graduate work at University of California, Berkeley, where he received his doctorate. His thesis dealt with the complex relationships and ecology of *Dichelostemma*, a splinter group of the native corm-bearing plants called brodiaeas.

His first teaching job took him to Durango in the Colorado Rockies, where he taught at Fort Lewis College for six years. Upon returning to California, he decided to pursue teaching adult education courses and has been doing so ever since. He has taught courses in botany and led field trips for the California Academy of Sciences, University of California Botanical Gardens, Yosemite Association, Mount Diablo Interpretive Association, and Strybing Arboretum. He served as education director at the latter establishment for six years and oversaw the docent programs there.

Glenn has always been interested in writing. His first efforts were the pocket guides in the finder series from Nature Study Guild: *Pacific Coast Berry Finder, Sierra Flower Finder,* and *Pacific Coast Fern Finder.* This led to more ambitious efforts, which have culminated in a pair of gardening books for Chronicle Books: *Complete Garden Guide to Native Perennials of California* and *Complete Garden Guide to Native Shrubs of California.*

Currently, Glenn lives on an acre in Sebastopol, where he tends his garden when time allows, and continues to teach and lead field trips. He hopes to do more writing for Mount Diablo Interpretive Association.

Peg Steunenberg received a Bachelor of Science Degree from Avila College, Kansas City, Missouri and later a Bachelor of Art Degree (Drawing and Painting) from San Jose State University. Additional art education included a one-year course of study in a government-sponsored Graphic Arts program. Scholarship funding was awarded for continuing study in illustration technique at the Academy of Art in San Francisco in conjunction with an internship at the Design Department, KQED-TV. Involvement with Mt. Diablo Interpretive Association and an annual educational product line provide interpretive information about the plants and animals of California to the public. At the time of this book's publication, Peg is a participant in the graduate studies program for Scientific Illustration at the University of California, Santa Cruz.

Susan Bazell has worked most of her adult life in fields unrelated to art or botany. Her love of the outdoors and wild plants dates from childhood and has recently led to her pursuit of drawing. She has taken several extension classes in botanical and scientific illustration at the California College of Arts and Crafts in Oakland, the University of California, Berkeley, and California State University, Hayward.

To order additional copies of this book or for more information about Mount Diablo State Park and other parks of the East Bay or the local organizations that have funded this project, please contact:

The Mount Diablo Interpretive Association
P.O. Box 346
Walnut Creek, CA 94597-0346
(510) 933-5289

Formed in 1974 as a state park cooperating association to educate the
public about the park's natural and cultural history and foster enlightened
use of this important resource.

Save Mount Diablo
P.O. Box 5376
Walnut Creek, CA 94596
(510) 229-4275

Formed in 1971 to acquire land and preserve Mount Diablo.

East Bay Regional Park District
2950 Peralta Oaks Court
P.O. Box 5381
Oakland, CA 94605-0381
(510) 653-0138, ext. 2203

California Native Plant Society
East Bay Chapter
P.O. Box 5597, Elmwood Station
Berkeley, CA 94705

Orinda Garden Club
P.O. Box 34
Orinda, CA 94563